Annals of Mathematics Studies

Number 180

CONVOLUTION AND EQUIDISTRIBUTION
Sato-Tate Theorems for Finite-Field Mellin Transforms

Nicholas M. Katz

PRINCETON UNIVERSITY PRESS
PRINCETON AND OXFORD
2012

Published by Princeton University Press, 41 William Street,
Princeton, New Jersey 08540

In the United Kingdom: Princeton University Press, 6 Oxford Street,
Woodstock, Oxfordshire OX20 1TW

Library of Congress Cataloging-in-Publication Data

Katz, Nicholas M., 1943-
 Convolution and equidistribution : Sato-Tate theorems for finite-field
Mellin transforms / Nicholas M. Katz.
  p. cm.
 Includes bibliographical references and index.
 ISBN 978-0-691-15330-8 (hardcover : alk. paper) – ISBN 978-0-691-
15331-5 (pbk. : alk. paper) 1. Mellin transform. 2. Convolutions
(Mathematics) 3. Sequences (Mathematics) I. Title.
 QA432.K38 2012
 515′.723–dc23   2011030224

British Library Cataloging-in-Publication Data is available

This book has been composed in LATEX.
The publisher would like to acknowledge the author of this
volume for providing the camera-ready copy from which this
book was printed.

Printed on acid-free paper ∞
press.princeton.edu
Printed in the United States of America

1 3 5 7 9 10 8 6 4 2

Contents

Convolution and Equidistribution

Introduction

The systematic study of character sums over finite fields may be said to have begun over 200 years ago, with Gauss. The Gauss sums over $\mathbb{F}_p$ are the sums

$$\sum_{x \in \mathbb{F}_p^\times} \psi(x)\chi(x),$$

for ψ a nontrivial additive character of $\mathbb{F}_p$, e.g., $x \mapsto e^{2\pi i x/p}$, and χ a nontrivial multiplicative character of $\mathbb{F}_p^\times$. Each has absolute value $\sqrt{p}$. In 1926, Kloosterman [**Kloos**] introduced the sums (one for each $a \in \mathbb{F}_p^\times$)

$$\sum_{xy=a \text{ in } \mathbb{F}_p} \psi(x+y)$$

which bear his name, in applying the circle method to the problem of four squares. In 1931 Davenport [**Dav**] became interested in (variants of) the following questions: for how many x in the interval $[1, p-2]$ are both x and $x+1$ squares in $\mathbb{F}_p$? Is the answer approximately $p/4$ as p grows? For how many x in $[1, p-3]$ are each of $x, x+1, x+2$ squares in $\mathbb{F}_p$? Is the answer approximately $p/8$ as p grows? For a fixed integer $r \geq 2$, and a (large) prime p, for how many x in $[1, p-r]$ are each of $x, x+1, x+2, ..., x+r-1$ squares in $\mathbb{F}_p$. Is the answer approximately $p/2^r$ as p grows? These questions led him to the problem of giving good estimates for character sums over the prime field $\mathbb{F}_p$ of the form

$$\sum_{x \in \mathbb{F}_p} \chi_2(f(x)),$$

where χ_2 is the quadratic character $\chi_2(x) := \left(\frac{x}{p}\right)$, and where $f(x) \in \mathbb{F}_p[x]$ is a polynomial with all distinct roots. Such a sum is the "error term" in the approximation of the number of mod p solutions of the equation

$$y^2 = f(x)$$

by p, indeed the number of mod p solutions is exactly equal to

$$p + \sum_{x \in \mathbb{F}_p} \chi_2(f(x)).$$

And, if one replaces the quadratic character by a character χ of $\mathbb{F}_p^\times$ of higher order, say order n, then one is asking about the number of mod p solutions of the equation

$$y^n = f(x).$$

This number is exactly equal to

$$p + \sum_{\chi | \chi^n = 1, \chi \neq 1} \sum_{x \in \mathbb{F}_p} \chi(f(x)).$$

The "right" bounds for Kloosterman's sums are

$$|\sum_{xy=a \text{ in } \mathbb{F}_p} \psi(x+y)| \leq 2\sqrt{p}$$

for $a \in \mathbb{F}_p^\times$. For $f(x) = \sum_{i=0}^d a_i x^i$ squarefree of degree d, the "right" bounds are

$$|\sum_{x \in \mathbb{F}_p} \chi(f(x))| \leq (d-1)\sqrt{p}$$

for χ nontrivial and $\chi^d \neq 1$, and

$$|\chi(a_d) + \sum_{x \in \mathbb{F}_p} \chi(f(x))| \leq (d-2)\sqrt{p}$$

for χ nontrivial and $\chi^d = 1$. These bounds were foreseen by Hasse [**Ha-Rel**] to follow from the Riemann Hypothesis for curves over finite fields, and were thus established by Weil [**Weil**] in 1948.

Following Weil's work, it is natural to "normalize" such a sum by dividing it by $\sqrt{p}$, and then ask how it varies in an algebro-geometric family. For example, one might ask how the normalized[1] Kloosterman sums

$$-(1/\sqrt{p}) \sum_{xy=a \text{ in } \mathbb{F}_p} \psi(x+y)$$

vary with $a \in \mathbb{F}_p^\times$, or how the sums

$$-(1/\sqrt{p}) \sum_{x \in \mathbb{F}_p} \chi_2(f(x))$$

vary as f runs over all squarefree cubic polynomials in $\mathbb{F}_p[x]$. [In this second case, we are looking at the $\mathbb{F}_p$-point count for the elliptic curve $y^2 = f(x)$.] Both these sorts of normalized sums are real, and lie in the closed interval $[-2, 2]$, so each can be written as twice the cosine of a

---

[1]The reason for introducing the minus sign will become clear later.

unique angle in $[0, \pi]$. Thus we define angles $\theta_{a,p}$, $a \in \mathbb{F}_p^\times$, and angles $\theta_{f,p}$, f a squarefree cubic in $\mathbb{F}_p[x]$:

$$-(1/\sqrt{p}) \sum_{xy=a \, in \, \mathbb{F}_p} \psi(x+y) = 2\cos\theta_{a,p},$$

$$-(1/\sqrt{p}) \sum_{x \in \mathbb{F}_p} \chi_2(f(x)) = 2\cos\theta_{f,p}.$$

In both these cases, the Sato-Tate conjecture asserted that, as p grows, the sets of angles $\{\theta_{a,p}\}_{a \, \in \, \mathbb{F}_p^\times}$ (respectively $\{\theta_{f,p}\}_{f \, \in \, \mathbb{F}_p[x] \, \text{squarefree cubic}}$) become equidistributed in $[0, \pi]$ for the measure $(2/\pi)\sin^2(\theta)d\theta$. Equivalently, the normalized sums themselves become equidistributed in $[-2, 2]$ for the "semicircle measure" $(1/2\pi)\sqrt{4-x^2}dx$. These Sato-Tate conjectures were shown by Deligne to fall under the umbrella of his general equidistribution theorem, cf. [**De-Weil II**, 3.5.3 and 3.5.7] and [**Ka-GKM**, 3.6 and 13.6]. Thus for example one has, for a fixed nontrivial χ, and a fixed integer $d \geq 3$ such that $\chi^d \neq \mathbb{1}$, a good understanding of the equidistribution properties of the sums

$$-(1/\sqrt{p}) \sum_{x \in \mathbb{F}_p} \chi(f(x))$$

as f ranges over various algebro-geometric families of polynomials of degree d, cf. [**Ka-ACT**, 5.13].

In this work, we will be interested in questions of the following type: fix a polynomial $f(x) \in \mathbb{F}_p[x]$, say squarefree of degree $d \geq 2$. For each multiplicative character χ with $\chi^d \neq \mathbb{1}$, we have the normalized sum

$$-(1/\sqrt{p}) \sum_{x \in \mathbb{F}_p} \chi(f(x)).$$

How are these normalized sums distributed as we **keep f fixed but vary χ** over all multiplicative characters χ with $\chi^d \neq \mathbb{1}$? More generally, suppose we are given some suitably algebro-geometric function $g(x)$, what can we say about suitable normalizations of the sums

$$\sum_{x \in \mathbb{F}_p} \chi(x)g(x)$$

as χ varies? This case includes the sums $\sum_{x \in \mathbb{F}_p} \chi(f(x))$, by taking for g the function $x \mapsto -1 + \#\{t \in \mathbb{F}_p | f(t) = x\}$, cf. Remark 17.7.

The earliest example we know in which this sort of question of variable χ is addressed is the case in which $g(x)$ is taken to be $\psi(x)$, so that we are asking about the distribution on the unit circle S^1 of the $p-2$ normalized Gauss sums

$$-(1/\sqrt{p}) \sum_{x \in \mathbb{F}_p^\times} \psi(x)\chi(x),$$

as χ ranges over the nontrivial multiplicative characters. The answer is that as p grows, these $p-2$ normalized sums become more and more equidistributed for Haar measure of total mass one in S^1. This results [**Ka-SE**, 1.3.3.1] from Deligne's estimate [**De-ST**, 7.1.3, 7.4] for multi-variable Kloosterman sums. There were later results [**Ka-GKM**, 9.3, 9.5] about equidistribution of r-tuples of normalized Gauss sums in $(S^1)^r$ for any $r \geq 1$. The theory we will develop here "explains" these last results in a quite satisfactory way, cf. Corollary 20.2.

Most of our attention is focused on equidistribution results over larger and larger finite extensions of a given finite field. Emanuel Kowalski drew our attention to the interest of having equidistribution results over, say, prime fields $\mathbb{F}_p$, that become better and better as p grows. This question is addressed in Chapter 28, where the problem is to make effective the estimates, already given in the equicharacteristic setting of larger and larger extensions of a given finite field. In Chapter 29, we point out some open questions about "the situation over $\mathbb{Z}$" and give some illustrative examples.

We end this introduction by pointing out two potential ambiguities of notation.

(1) We will deal both with lisse sheaves, usually denoted by calligraphic letters, most commonly $\mathcal{F}$, on open sets of $\mathbb{G}_m$, and with perverse sheaves, typically denoted by roman letters, most commonly N and M, on $\mathbb{G}_m$. We will develop a theory of the Tannakian groups $G_{geom,N}$ and $G_{arith,N}$ attached to (suitable) perverse sheaves N. We will also on occasion, especially in Chapters 11 and 12, make use of the "usual" geometric and arithmetic monodromy groups $G_{geom,\mathcal{F}}$ and $G_{arith,\mathcal{F}}$ attached to lisse sheaves $\mathcal{F}$. The difference in typography, which in turns indicates whether one is dealing with a perverse sheaf or a lisse sheaf, should always make clear which sort of G_{geom} or G_{arith} group, the Tannakian one or the "usual" one, is intended.

(2) When we have a lisse sheaf $\mathcal{F}$ on an open set of $\mathbb{G}_m$, we often need to discuss the representation of the inertia group $I(0)$ at 0 (respectively the representation of the inertia group $I(\infty)$ at ∞) to which $\mathcal{F}$ gives rise. We will denote these representations $\mathcal{F}(0)$ and $\mathcal{F}(\infty)$ respectively. We will also wish to consider Tate twists $\mathcal{F}(n)$ or $\mathcal{F}(n/2)$ of $\mathcal{F}$ by **nonzero** integers n or half-integers $n/2$. We adopt the convention that $\mathcal{F}(0)$ (or $\mathcal{F}(\infty)$) always means the representation of the

corresponding inertia group, while $\mathcal{F}(n)$ or $\mathcal{F}(n/2)$ with n a nonzero integer always means a Tate twist.

CHAPTER 1

Overview

Let k be a finite field, q its cardinality, p its characteristic,

$$\psi : (k, +) \to \mathbb{Z}[\zeta_p]^\times \subset \mathbb{C}^\times$$

a nontrivial additive character of k, and

$$\chi : (k^\times, \times) \to \mathbb{Z}[\zeta_{q-1}]^\times \subset \mathbb{C}^\times$$

a (possibly trivial) multiplicative character of k.

The present work grew out of two questions, raised by Ron Evans and Zeev Rudnick respectively, in May and June of 2003. Evans had done numerical experiments on the sums

$$S(\chi) := -(1/\sqrt{q}) \sum_{t \in k^\times} \psi(t - 1/t)\chi(t)$$

as χ varies over all multiplicative characters of k. For each χ, $S(\chi)$ is real, and (by Weil) has absolute value at most 2. Evans found empirically that, for large $q = \#k$, these $q - 1$ sums were approximately equidistributed for the "Sato-Tate measure"[1] $(1/2\pi)\sqrt{4 - x^2}dx$ on the closed interval $[-2, 2]$, and asked if this equidistribution could be proven.

Rudnick had done numerical experiments on the sums

$$T(\chi) := -(1/\sqrt{q}) \sum_{t \in k^\times, t \neq 1} \psi((t + 1)/(t - 1))\chi(t)$$

as χ varies now over all *nontrivial* multiplicative characters of a finite field k of *odd* characteristic, cf. [**KRR**, Appendix A] for how these sums arose. For nontrivial χ, $T(\chi)$ is real, and (again by Weil) has absolute value at most 2. Rudnick found empirically that, for large $q = \#k$, these $q-2$ sums were approximately equidistributed for the same "Sato-Tate measure" $(1/2\pi)\sqrt{4 - x^2}dx$ on the closed interval $[-2, 2]$, and asked if this equidistribution could be proven.

---

[1]This is the measure which is the direct image of the total mass one Haar measure on the compact group $SU(2)$ by the trace map Trace : $SU(2) \to [-2, 2]$, i.e., it is the measure according to which traces of "random" elements of $SU(2)$ are distributed.

We will prove both of these equidistribution results. Let us begin by slightly recasting the original questions. Fixing the characteristic p of k, we choose a prime number $\ell \neq p$; we will soon make use of ℓ-adic étale cohomology. We denote by $\mathbb{Z}_\ell$ the ℓ-adic completion of $\mathbb{Z}$, by $\mathbb{Q}_\ell$ its fraction field, and by $\overline{\mathbb{Q}}_\ell$ an algebraic closure of $\mathbb{Q}_\ell$. We also choose a field embedding ι of $\overline{\mathbb{Q}}_\ell$ into $\mathbb{C}$. Any such ι induces an isomorphism between the algebraic closures of $\mathbb{Q}$ in $\overline{\mathbb{Q}}_\ell$ and in $\mathbb{C}$ respectively.[2] By means of ι, we can, on the one hand, view the sums $S(\chi)$ and $T(\chi)$ as lying in $\overline{\mathbb{Q}}_\ell$. On the other hand, given an element of $\overline{\mathbb{Q}}_\ell$, we can ask if it is real, and we can speak of its complex absolute value. This allows us to define what it means for a lisse sheaf to be ι-pure of some weight w (and later, for a perverse sheaf to be ι-pure of some weight w). We say that a perverse sheaf is pure of weight w if it is ι-pure of weight w for every choice of ι.

By means of the chosen ι, we view both the nontrivial additive character ψ of k and every (possibly trivial) multiplicative character χ of $k^\times$ as having values in $\overline{\mathbb{Q}}_\ell^\times$. Then, attached to ψ, we have the Artin-Schreier sheaf $\mathcal{L}_\psi = \mathcal{L}_{\psi(x)}$ on $\mathbb{A}^1/k := Spec(k[x])$, a lisse $\overline{\mathbb{Q}}_\ell$-sheaf of rank one on $\mathbb{A}^1/k$ which is pure of weight zero. And for each χ we have the Kummer sheaf $\mathcal{L}_\chi = \mathcal{L}_{\chi(x)}$ on $\mathbb{G}_m/k := Spec(k[x, 1/x])$, a lisse $\overline{\mathbb{Q}}_\ell$-sheaf of rank one on $\mathbb{G}_m/k$ which is pure of weight zero. For a k-scheme X and a k-morphism $f : X \to \mathbb{A}^1/k$ (resp. $f : X \to \mathbb{G}_m/k$), we denote by $\mathcal{L}_{\psi(f)}$ (resp. $\mathcal{L}_{\chi(f)}$) the pullback lisse rank one, pure of weight zero, sheaf $f^\star \mathcal{L}_{\psi(x)}$ (resp. $f^\star \mathcal{L}_{\chi(x)}$) on X.

In the question of Evans, we view $x - 1/x$ as a morphism from $\mathbb{G}_m$ to $\mathbb{A}^1$, and form the lisse sheaf $\mathcal{L}_{\psi(x-1/x)}$ on $\mathbb{G}_m/k$. In the question of Rudnick, we view $(x+1)/(x-1)$ as a morphism from $\mathbb{G}_m \setminus \{1\}$ to $\mathbb{A}^1$, and form the lisse sheaf $\mathcal{L}_{\psi((x+1)/(x-1))}$ on $\mathbb{G}_m \setminus \{1\}$. With

$$j : \mathbb{G}_m \setminus \{1\} \to \mathbb{G}_m$$

the inclusion, we form the direct image sheaf $j_\star \mathcal{L}_{\psi((x+1)/(x-1))}$ on $\mathbb{G}_m/k$ (which for this sheaf, which is totally ramified at the point 1, is the same as extending it by zero across the point 1).

The common feature of both questions is that we have a dense open set $U/k \subset \mathbb{G}_m/k$, a lisse, ι-pure of weight zero sheaf $\mathcal{F}$ on U/k, its extension $\mathcal{G} := j_\star \mathcal{F}$ by direct image to $\mathbb{G}_m/k$, and we are looking at the sums

$$-(1/\sqrt{q}) \sum_{t \in \mathbb{G}_m(k) = k^\times} \chi(t) \text{Trace}(Frob_{t,k} | \mathcal{G})$$

---

[2] Such an ι need not be a field isomorphism of $\overline{\mathbb{Q}}_\ell$ with $\mathbb{C}$, but we may choose an ι which is, as Deligne did in [**De-Weil II**, 0.2, 1.2.6, 1.2.11].

$$= -(1/\sqrt{q}) \sum_{t \in \mathbb{G}_m(k)=k^\times} \mathrm{Trace}(Frob_{t,k}|\mathcal{G} \otimes \mathcal{L}_\chi).$$

To deal with the factor $1/\sqrt{q}$, we choose a square root of the ℓ-adic unit p in $\overline{\mathbb{Q}}_\ell$, and use powers of this chosen square root as our choices of $\sqrt{q}$. [For definiteness, we might choose that $\sqrt{p}$ which via ι becomes the positive square root, but either choice will do.] Because $\sqrt{q}$ is an ℓ-adic unit, we may form the "half"-Tate twist $\mathcal{G}(1/2)$ of $\mathcal{G}$, which for any finite extension field E/k and any point $t \in \mathbb{G}_m(E)$ multiplies the traces of the Frobenii by $1/\sqrt{\#E}$, i.e.,

$$\mathrm{Trace}(Frob_{t,E}|\mathcal{G}(1/2)) = (1/\sqrt{\#E})\mathrm{Trace}(Frob_{t,E}|\mathcal{G}).$$

As a final and apparently technical step, we replace the middle extension sheaf $\mathcal{G}(1/2)$ by the same sheaf, but now placed in degree -1, namely the object

$$M := \mathcal{G}(1/2)[1]$$

in the derived category $D_c^b(\mathbb{G}_m/k, \overline{\mathbb{Q}}_\ell)$. It will be essential in a moment that the object M is in fact a perverse sheaf, but for now we need observe only that this shift by one of the degree has the effect of changing the sign of each Trace term. In terms of this object, we are looking at the sums

$$S(M, k, \chi) := \sum_{t \in \mathbb{G}_m(k)=k^\times} \chi(t)\mathrm{Trace}(Frob_{t,k}|M).$$

So written, the sums $S(M, k, \chi)$ make sense for *any* object $M \in D_c^b(\mathbb{G}_m/k, \overline{\mathbb{Q}}_\ell)$. If we think of M as fixed but χ as variable, we are looking at the Mellin (:= multiplicative Fourier) transform of the function $t \mapsto \mathrm{Trace}(Frob_{t,k}|M)$ on the finite abelian group $\mathbb{G}_m(k) = k^\times$. It is a standard fact that the Mellin transform turns multiplicative convolution of functions on $k^\times$ into multiplication of functions of χ.

On the derived category $D_c^b(\mathbb{G}_m/k, \overline{\mathbb{Q}}_\ell)$, we have a natural operation of !-convolution

$$(M, N) \to M \star_! N$$

defined in terms of the multiplication map

$$\pi : \mathbb{G}_m \times \mathbb{G}_m \to \mathbb{G}_m, \quad (x, y) \to xy$$

and the external tensor product object

$$M \boxtimes N := pr_1^\star M \otimes pr_2^\star M$$

in $D_c^b(\mathbb{G}_m \times \mathbb{G}_m/k, \overline{\mathbb{Q}}_\ell)$ as

$$M \star_! N := R\pi_!(M \boxtimes N).$$

It then results from the Lefschetz Trace formula [**Gr-Rat**] and proper base change that, for any multiplicative character χ of $k^\times$, we have the product formula

$$S(M \star_! N, k, \chi) = S(M, k, \chi)S(N, k, \chi);$$

more generally, for any finite extension field E/k, and any multiplicative character ρ of $E^\times$, we have the product formula

$$S(M \star_! N, E, \rho) = S(M, E, \rho)S(N, E, \rho).$$

At this point, we must mention two technical points, which will be explained in detail in the next chapter, but which we will admit here as black boxes. The first is that we must work with perverse sheaves N satisfying a certain supplementary condition, $\mathcal{P}$. This is the condition that, working on $\mathbb{G}_m/\overline{k}$, N admits no subobject and no quotient object which is a (shifted) Kummer sheaf $\mathcal{L}_\chi[1]$. For an N which is geometrically irreducible, $\mathcal{P}$ is simply the condition that N is not geometrically a (shifted) Kummer sheaf $\mathcal{L}_\chi[1]$. Thus any geometrically irreducible N which has generic rank ≥ 2, or which is not lisse on $\mathbb{G}_m$, or which is not tamely ramified at both 0 and ∞, certainly satisfies $\mathcal{P}$. Thus for example the object giving rise to the Evans sums, namely $\mathcal{L}_{\psi(x-1/x)}(1/2)[1]$, is wildly ramified at both 0 and ∞, and the object giving rise to the Rudnick sums, namely $j_\star \mathcal{L}_{\psi((x+1)/(x-1))}(1/2)[1]$, is not lisse at $1 \in \mathbb{G}_m(\overline{k})$, so both these objects satisfy $\mathcal{P}$. The second technical point is that we must work with a variant of ! convolution $\star_!$, called "middle" convolution $\star_{mid}$, which is defined on perverse sheaves satisfying $\mathcal{P}$, cf. the next chapter.

In order to explain the simple underlying ideas, we will admit four statements, and explain how to deduce from them equidistribution theorems about the sums $S(M, k, \chi)$ as χ varies.

(1) If M and N are both perverse on $\mathbb{G}_m/k$ (resp. on $\mathbb{G}_m/\overline{k}$) and satisfy $\mathcal{P}$, then their middle convolution $M \star_{mid} N$ is perverse on $\mathbb{G}_m/k$ (resp. on $\mathbb{G}_m/\overline{k}$) and satisfies $\mathcal{P}$.

(2) With the operation of middle convolution as the "tensor product," the skyscraper sheaf δ_1 as the "identity object," and $[x \mapsto 1/x]^\star DM$ as the "dual" $M^\vee$ of M (DM denoting the Verdier dual of M), the category of perverse sheaves on $\mathbb{G}_m/k$ (resp. on $\mathbb{G}_m/\overline{k}$) satisfying $\mathcal{P}$ is a neutral Tannakian category, in which the "dimension" of an object M is its Euler characteristic $\chi_c(\mathbb{G}_m/\overline{k}, M)$.

(3) Denoting by

$$j_0 : \mathbb{G}_m/\overline{k} \subset \mathbb{A}^1/\overline{k}$$

the inclusion, the construction

$$M \mapsto H^0(\mathbb{A}^1/\overline{k}, j_{0!}M)$$

is a fibre functor on the Tannakian category of perverse sheaves on $\mathbb{G}_m/\overline{k}$ satisfying $\mathcal{P}$ (and hence also a fibre functor on the subcategory of perverse sheaves on $\mathbb{G}_m/k$ satisfying $\mathcal{P}$). For $i \neq 0$, $H^i(\mathbb{A}^1/\overline{k}, j_{0!}M)$ vanishes.

(4) For any finite extension field E/k, and any multiplicative character ρ of $E^\times$, the construction

$$M \mapsto H^0(\mathbb{A}^1/\overline{k}, j_{0!}(M \otimes \mathcal{L}_\rho))$$

is also such a fibre functor. For $i \neq 0$, $H^i(\mathbb{A}^1/\overline{k}, j_{0!}(M \otimes \mathcal{L}_\rho))$ vanishes.

Now we make use of these four statements. Take for N a perverse sheaf on $\mathbb{G}_m/k$ which is ι-pure of weight zero and which satisfies $\mathcal{P}$. Denote by $<N>_{arith}$ the full subcategory of all perverse sheaves on $\mathbb{G}_m/k$ consisting of all subquotients of all "tensor products" of copies of N and its dual $N^\vee$. Similarly, denote by $<N>_{geom}$ the full subcategory of all perverse sheaves on $\mathbb{G}_m/\overline{k}$ consisting of all subquotients, in this larger category, of all "tensor products" of copies of N and its dual $N^\vee$. With respect to a choice ω of fibre functor, the category $<N>_{arith}$ becomes the category of finite-dimensional $\overline{\mathbb{Q}_\ell}$-representations of an algebraic group $G_{arith,N,\omega} \subset GL(\omega(N)) = GL(\text{"}\dim\text{"}N)$, with N itself corresponding to the given "$\dim$"N-dimensional representation. Concretely, $G_{arith,N,\omega} \subset GL(\omega(N))$ is the subgroup consisting of those automorphisms γ of $\omega(N)$ with the property that γ, acting on $\omega(M)$, for M any tensor construction on $\omega(N)$ and its dual, maps to itself every vector space subquotient of the form $\omega(\text{any subquotient of M})$.

And the category $<N>_{geom}$ becomes the category of finite-dimensional $\overline{\mathbb{Q}_\ell}$-representations of a possibly smaller algebraic group $G_{geom,N,\omega} \subset G_{arith,N,\omega}$ (smaller because there are more subobjects to be respected).

For ρ a multiplicative character of a finite extension field E/k, we have the fibre functor ω_ρ defined by

$$M \mapsto H^0(\mathbb{A}^1/\overline{k}, j_!(M \otimes \mathcal{L}_\rho))$$

on $<N>_{arith}$. The Frobenius $Frob_E$ is an automorphism of this fibre functor, so defines an **element** $Frob_{E,\rho}$ in the group G_{arith,N,ω_ρ} defined by this choice of fibre functor. But one knows that the groups $G_{arith,N,\omega}$ (respectively the groups $G_{geom,N,\omega}$) defined by different fibre functors are pairwise isomorphic, by a system of isomorphisms which are unique up to inner automorphism of source (or target). Fix one choice, say

ω_0, of fibre functor, and define

$$G_{arith,N} := G_{arith,N,\omega_0}, \quad G_{geom,N} := G_{geom,N,\omega_0}.$$

Then the **element** $Frob_{E,\rho}$ in the group G_{arith,N,ω_ρ} still makes sense as a **conjugacy class** in the group $G_{arith,N}$.

Let us say that a multiplicative character ρ of some finite extension field E/k is good for N if, for

$$j : \mathbb{G}_m/\overline{k} \subset \mathbb{P}^1/\overline{k}$$

the inclusion, the canonical "forget supports" map

$$Rj_!(N \otimes \mathcal{L}_\rho) \to Rj_\star(N \otimes \mathcal{L}_\rho)$$

is an isomorphism. If ρ is good for N, then the natural "forget supports" maps

$$H^0_c(\mathbb{G}_m/\overline{k}, N \otimes \mathcal{L}_\rho) = H^0_c(\mathbb{A}^1/\overline{k}, j_{0!}(N \otimes \mathcal{L}_\rho)) \to H^0(\mathbb{A}^1/\overline{k}, j_{0!}(N \otimes \mathcal{L}_\rho)),$$

together with the restriction map

$$H^0(\mathbb{A}^1/\overline{k}, j_{0!}(N \otimes \mathcal{L}_\rho)) \to H^0(\mathbb{G}_m/\overline{k}, N \otimes \mathcal{L}_\rho),$$

are all isomorphisms. Moreover, as N is ι-pure of weight zero, each of these groups is ι-pure of weight zero.

Conversely, if the group $\omega_\rho(N) := H^0(\mathbb{A}^1/\overline{k}, j_{0!}(N \otimes \mathcal{L}_\rho))$ is ι-pure of weight zero, then ρ is good for N, and we have a "forget supports" isomorphism

$$H^0_c(\mathbb{G}_m/\overline{k}, N \otimes \mathcal{L}_\rho) \cong \omega_\rho(N) := H^0(\mathbb{A}^1/\overline{k}, j_{0!}(N \otimes \mathcal{L}_\rho)).$$

This criterion, that ρ is good for N if and only if $\omega_\rho(N)$ is ι-pure of weight zero, shows that if ρ is good for N, then ρ is good for every object M in the Tannakian category $<N>_{arith}$ generated by N, and hence that for any such M, we have an isomorphism

$$H^0_c(\mathbb{G}_m/\overline{k}, M \otimes \mathcal{L}_\rho) \cong \omega_\rho(M).$$

Recall that geometrically, i.e., on $\mathbb{G}_m/\overline{k}$, we may view the various Kummer sheaves $\mathcal{L}_\rho$ coming from multiplicative characters ρ of finite subfields $E \subset \overline{k}$ as being the characters of finite order of the tame inertia group $I(0)^{tame}$ at 0, or of the tame inertia group $I(\infty)^{tame}$ at ∞, or of the tame fundamental group $\pi_1^{tame}(\mathbb{G}_m/\overline{k})$. In this identification, given a character ρ of a finite extension E/k and a further finite extension L/E, the pair (E, ρ) and the pair $(L, \rho \circ Norm_{L/E})$ give rise to the same Kummer sheaf on $\mathbb{G}_m/\overline{k}$. Up to this identification of (E, ρ) with $(L, \rho \circ Norm_{L/E})$, there are, for a given N, at most finitely many ρ which fail to be good for N (simply because there are at most finitely many tame characters which occur in the local monodromies of N at either 0 or ∞, and we need only avoid their inverses). Indeed, if we denote by $rk(N)$ the generic rank of N, there are at most $2rk(N)$ bad ρ for N.

Recall [**BBD**, 5.3.8] that a perverse N which is ι-pure of weight zero is geometrically semisimple. View N as a faithful representation of $G_{geom,N}$. Then $G_{geom,N}$ has a faithful, completely reducible representation, hence $G_{geom,N}$ is a reductive group.

Let us now suppose further that N is, in addition, arithmetically semisimple (e.g., arithmetically irreducible). Then $G_{arith,N}$ is also a reductive group. Choose a maximal compact subgroup K of the reductive Lie group $G_{arith,N}(\mathbb{C})$ (where we use ι to view $G_{arith,N}$ as an algebraic group over $\mathbb{C}$). For each finite extension field E/k and each character ρ of $E^\times$ which is good for N, we obtain a Frobenius conjugacy class $\theta_{E,\rho}$ in K as follows. Because ρ is good for N, $Frob_E$ has, via ι, unitary eigenvalues acting on $\omega_\rho(N)$, i.e., the conjugacy class $Frob_{E,\rho}$ in $G_{arith,N}$ has unitary eigenvalues when viewed in the ambient $GL(\omega_0(N))$. Therefore its semisimplification in the sense of the Jordan decomposition, $Frob_{E,\rho}^{ss}$, is a semisimple class in $G_{arith,N}(\mathbb{C})$ with unitary eigenvalues. Therefore any element in the class $Frob_{E,\rho}^{ss}$ lies in a compact subgroup of $G_{arith,N}(\mathbb{C})$ (e.g., in the closure of the subgroup it generates), and hence lies in a maximal compact subgroup of $G_{arith,N}(\mathbb{C})$. All such are $G_{arith,N}(\mathbb{C})$-conjugate, so we conclude that every element in the class $Frob_{E,\rho}^{ss}$ is conjugate to an element of K. We claim that this element is in turn well-defined in K up to K-conjugacy, so gives us a K-conjugacy class $\theta_{E,\rho}$. To show that $\theta_{E,\rho}$ is well-defined up to K-conjugacy, it suffices, by Peter-Weyl, to specify its trace in every finite-dimensional, continuous, unitary representation Λ_K of K. By Weyl's unitarian trick, every Λ_K of K is the restriction to K of a unique finite-dimensional representation Λ of the $\mathbb{C}$-group $G_{arith,N}/\mathbb{C}$. Thus for every Λ_K, we have the identity

$$\text{Trace}(\Lambda_K(\theta_{E,\rho})) = \text{Trace}(\Lambda(Frob\theta_{E,\rho}^{ss})) = \text{Trace}(\Lambda(Frob\theta_{E,\rho})).$$

With these preliminaries out of the way, we can state the main theorem.

Theorem 1.1. *Let N be an arithmetically semisimple perverse sheaf on $\mathbb{G}_m/k$ which is ι-pure of weight zero and which satisfies condition $\mathcal{P}$. Choose a maximal compact subgroup K of the reductive Lie group $G_{arith,N}(\mathbb{C})$. Suppose that we have an **equality** of groups*

$$G_{geom,N} = G_{arith,N}.$$

Then as E/k runs over larger and larger finite extension fields, the conjugacy classes $\{\theta_{E,\rho}\}_{good\ \rho}$ become equidistributed in the space $K^{\#}$

of conjugacy classes in K, for the induced Haar measure of total mass one.

Proof. For each finite extension field E/k, denote by $\mathrm{Good}(E, N)$ the set of multiplicative characters ρ of $E^\times$ which are good for N. We must show that for any continuous central function f on K, we can compute $\int_K f(k)dk$ as the limit, as E/k runs over larger and larger extension fields, large enough that the set Good(E, N) is nonempty, of the sums

$$(1/\#\mathrm{Good}(E, N)) \sum_{\rho \in Good(E,N)} f(\theta_{E,\rho}).$$

By the Peter-Weyl theorem, any such f is a uniform limit of finite linear combinations of traces of irreducible representations of K. So it suffices to check when f is the trace of an irreducible representation Λ_K of K. If Λ_K is the trivial representation, both the integral and each of the sums is identically 1.

For an irreducible nontrivial representation Λ_K of K, corresponding to an irreducible nontrivial representation Λ of $G_{arith,N}$, we have

$$\int_K \mathrm{Trace}(\Lambda_K(k))dk = 0.$$

So it remains to show that for any such Λ_K, as $\#E \to \infty$ over finite extensions large enough that $\mathrm{Good}(E, N)$ is nonempty, we have

$$(1/\#\mathrm{Good}(E, N)) \sum_{\rho \in Good(E,N)} \mathrm{Trace}(\Lambda_K(\theta_{E,\rho}))$$

$$\to \int_K \mathrm{Trace}(\Lambda_K(k))dk = 0.$$

To see that this is the case, denote by M the object corresponding to the irreducible nontrivial representation Λ of $G_{arith,N}$. By the hypothesis that $G_{geom,N} = G_{arith,N}$, it follows that M is itself geometrically irreducible and nontrivial. In terms of this object, the sums above are the sums

$$(1/\#\mathrm{Good}(E, N)) \sum_{\rho \in Good(E,N)} S(M, E, \rho).$$

We next compare these sums to the sums

$$(1/\#E^\times) \sum_{\rho \text{ char of } E^\times} S(M, E, \rho).$$

Each individual term $S(M, E, \rho)$ has $|S(M, E, \rho)| \le$ "dim"M (cf. Lemmas 2.1 and 3.4), and the number of terms in the two sums compare as

$$\#E^\times \ge \#\mathrm{Good}(E, N) \ge \#E^\times - 2rk(N).$$

So it is equivalent to show instead that the sums

$$(1/\#E^\times) \sum_{\rho \ char \ of \ E^\times} S(M,E,\rho)$$

tend to 0 as $\#E$ grows.

We now distinguish two cases. If M is punctual, then it must be a (unitary, but we will not need this) constant field twist $\alpha^{deg}\delta_{t_0}$ of a skyscraper sheaf δ_{t_0} concentrated at a point $t_0 \neq 1$ in $\mathbb{G}_m(k)$ ($\neq 1$ because geometrically nontrivial, a single rational point because punctual and both arithmetically and geometrically irreducible). In this case we have

$$S(M,E,\rho) = \alpha^{degE/k}\rho(t_0),$$

and the sum over all ρ vanishes (orthogonality of characters).

If M is nonpunctual, then it is of the form $\mathcal{G}(1/2)[1]$ for a geometrically irreducible middle extension sheaf $\mathcal{G}$ on $\mathbb{G}_m/k$ which is ι-pure of weight zero and which is not geometrically isomorphic to any shifted Kummer sheaf $\mathcal{L}_\chi[1]$. In this case, $rk(M) = rk(\mathcal{G})$, the generic rank of $\mathcal{G}$. Here

$$S(M,E,\rho) = (-1/\sqrt{\#E}) \sum_{t\in E^\times} \rho(t)\mathrm{Trace}(Frob_{E,t}|\mathcal{G}).$$

By orthogonality, we get

$$(1/\#E^\times) \sum_{\rho \ char \ of \ E^\times} S(M,E,\rho) = (-1/\sqrt{\#E})\mathrm{Trace}(Frob_{E,1}|\mathcal{G}).$$

As $\mathcal{G}$ is a middle extension which is ι-pure of weight zero, we have $|\mathrm{Trace}(Frob_{E,1}|\mathcal{G})| \leq rk(\mathcal{G})$, cf. [**De-Weil II**, 1.8.1]. $\square$

Taking the direct image of this approximate equidistribution of conjugacy classes by the Trace map, we get the following corollary, which addresses the sums with which we were originally concerned.

Corollary 1.2. *As E/k runs over larger and larger finite extension fields, the sums $\{S(N,E,\rho)\}_{good \ \rho}$ become equidistributed in $\mathbb{C}$ for the probability measure which is the direct image by the Trace map*

$$\mathrm{Trace} : K \to \mathbb{C}$$

of Haar measure of total mass one on K.

In the book itself, we will explain in greater detail both condition $\mathcal{P}$ and middle convolution. Here we owe a tremendous debt to the paper [**Ga-Loe**] of Gabber and Loeser for basic Tannakian facts, and to Deligne for suggesting the fibre functor of which we make essential use, cf. [**Ka-GKM**, 2.2.1] where we used Deligne's fibre functor in a slightly different context.

Once we have proper foundations, we then explore various cases where we can both show that $G_{geom,N} = G_{arith,N}$ and compute what this group is. For example, we will show that in both the Evans and Rudnick situations, we have $G_{geom,N} = G_{arith,N} = SL(2)$, thus $K = SU(2)$, and we recover the approximate equidistribution of their sums for the classical Sato-Tate measure on $[-2, 2]$. We will also give examples where the common value of the groups is $GL(n)$, any n, or $Sp(2n)$, any n, or $SO(n)$, any $n \geq 3$, or $SL(n)$, any $n \geq 3$ which is either a prime number[3] or not a power of the characteristic, or $O(2n)$, any $n \geq 2$ which is not a power of the characteristic. We also give, in every characteristic, families of objects "most of which" give rise to the exceptional group G_2, cf. Theorem 25.1. We do not know how to obtain the other exceptional groups.

For a broader perspective on the Tannakian approach we take here, consider the corresponding additive version of this same problem, where we are now given a geometrically irreducible perverse sheaf M on $\mathbb{A}^1/k$ which is pure of weight zero. For variable additive characters ρ of k, we wish to study the sums

$$S(M, k, \rho) := \sum_{x \in \mathbb{A}^1(k)} \rho(x) \text{Trace}(Frob_{x,k}|M).$$

In this additive case, we have the advantage that having fixed one choice of nontrivial additive character ψ, the additive characters of k are each of the form $x \mapsto \psi(tx)$ for a unique $t \in k$, so these sums are the sums

$$S(M, k, t) := \sum_{x \in \mathbb{A}^1(k)} \psi(tx) \text{Trace}(Frob_{x,k}|M).$$

These sums are (minus) the traces of Frobenius on the Fourier Transform of M, cf. [**Lau-TFCEF**, 1.2.1.2]. Recall that for any $N \in D_c^b(\mathbb{A}^1, \overline{\mathbb{Q}_\ell})$, its Fourier Transform $FT(N)$ is defined by

$$FT(N) := R(pr_2)_!(pr_1^\star N \otimes \mathcal{L}_{\psi(tx)}[1]),$$

where pr_1 and pr_2 are the two projections of $\mathbb{A}^2$, with coordinates (x, t), onto $\mathbb{A}^1$. The formation of the Fourier Transform is essentially involutive, and, by the "miracle" of Fourier Transform, if M is a geometrically irreducible perverse sheaf on $\mathbb{A}^1$ which is pure of weight zero, then $FT(M)$ is a geometrically irreducible perverse sheaf on $\mathbb{A}^1$ which is

---

[3] In the examples where we require that n be prime, this requirement is a reflection of our ignorance; see Chapter 21.

pure of weight one, cf. [**Ka-Lau**, 2.1.3, 2.1.5, 2.2.1] and [**Lau-TFCEF**, 1.2.2.3, 1.3.2.3, 1.3.2.4]. If in addition M is not geometrically isomorphic to a shifted Artin-Schreier sheaf $\mathcal{L}_{\psi(ax)}[1]$ for some a, then $FT(M)$ is of the form $\mathcal{G}[1]$, where $\mathcal{G}$ is a geometrically irreducible middle extension sheaf on $\mathbb{A}^1/k$ which is pure of weight zero, cf. [**Lau-TFCEF**, 1.4.2.1] and [**Ka-Lau**, 2.2.1]. The sums above are simply given by

$$\text{Trace}(Frob_{t,k}|\mathcal{G}) = \sum_{x\in\mathbb{A}^1(k)} \psi(tx)\text{Trace}(Frob_{x,k}|M),$$

cf. [**Lau-TFCEF**, 1.2.1.2]. So in this additive case, to study the equidistribution properties of these sums, we need "only" apply Deligne's equidistribution theorem to the sheaf $\mathcal{G}$, restricted to a dense open set, say U, where it is lisse. Then we are "reduced" to computing the Zariski closure of the geometric monodromy group of $\mathcal{G}|U$, and showing that the Zariski closure of the arithmetic monodromy group is no bigger, cf. [**Ka-GKM**, Chapter 3]. In many cases, the computation of the geometric monodromy group depends on knowing the local monodromies of $\mathcal{G}$ at the points of $\mathbb{P}^1 \setminus U$, where $\mathcal{G}$ fails to be lisse.

If we were to transpose to the additive case the techniques we develop here to handle the case of variable multiplicative characters χ, it would amount to the following. Another interpretation of the Zariski closure of the geometric (resp. arithmetic) monodromy group of $\mathcal{G}|U$ is that it is the Tannakian group governing the tensor category generated by $\mathcal{G}|U_{\overline{k}}$ (resp. by $\mathcal{G}|U$). This tensor category for $\mathcal{G}$, with usual tensor product as the tensor operation, is tensor equivalent to the Tannakian category generated by the input object M on $\mathbb{A}^1$, where now the "tensor operation" is additive middle convolution (reflecting the fact that on functions, Fourier Transform interchanges multiplication and convolution). So we would end up studying the Tannakian subcategory, generated by M, consisting of those perverse sheaves on $\mathbb{A}^1$ which satisfy the additive version of condition $\mathcal{P}$, under the "tensor" operation of additive middle convolution. In such a study, we do not "see" the local monodromies of $\mathcal{G}$, or even the existence of $\mathcal{G}$. So in the additive case, the methods we develop here would lead us to attempt to prove, in some few cases successfully, known theorems about the monodromy groups of various Fourier Transform sheaves with our hands tied behind our back.

In this work, dealing with varying multiplicative characters and an arithmetically semisimple perverse N on $\mathbb{G}_m/k$ which is pure of weight zero and satisfies $\mathcal{P}$, we have a weak substitute for the apparently nonexistent notion of "local monodromy" at a bad (for N) multiplicative character ρ of some finite extension E/k. This substitute, cf.

Chapter 16, is Serre's Frobenius torus attached to the (semisimplified) Frobenius class $Frob_{E,\rho}^{ss}$, cf. [**Se-Let**]. In fact, following an idea of Deligne [**Se-Rep**, 2.3], we make use only of the archimedean absolute values of the eigenvalues of $Frob_{E,\rho}$, and the fact that these absolute values are not all equal to one (because of the badness of ρ). This approach breaks down completely for an N which has no bad ρ, i.e., for an N which is totally wildly ramified at both 0 and ∞. Much remains to be done.

CHAPTER 2

Convolution of Perverse Sheaves

Let k be a finite field, q its cardinality, p its characteristic, $\ell \neq p$ a prime number, and G/k a smooth commutative groupscheme which over $\overline{k}$ becomes isomorphic to $\mathbb{G}_m/\overline{k}$. We will be concerned with perverse sheaves on G/k and on $G/\overline{k}$.

We begin with perverse sheaves on $G/\overline{k} \cong \mathbb{G}_m/\overline{k}$. On the derived category $D_c^b(\mathbb{G}_m/k, \overline{\mathbb{Q}}_\ell)$ we have two notions of convolution, ! convolution and $\star$ convolution, defined respectively by

$$N \star_! M := R\pi_!(N \boxtimes M),$$

$$N \star_\star M := R\pi_\star(N \boxtimes M),$$

where $\pi : \mathbb{G}_m \times \mathbb{G}_m \to \mathbb{G}_m$ is the multiplication map. For neither of these notions is it the case that the convolution of two perverse sheaves need be perverse. In our book [Ka-RLS], we addressed this difficulty by introducing the full subcategory $\mathcal{P}$ of all perverse sheaves consisting of those perverse sheaves N with the property that for any perverse sheaf M, both convolutions $N \star_! M$ and $N \star_\star M$ were perverse. For N and M both in $\mathcal{P}$, we then defined their middle convolution $N \star_{mid} M$ as

$$N \star_{mid} M := \mathrm{Image}(N \star_! M \to N \star_\star M)$$

under the natural "forget supports" map. We viewed the perverse sheaves with $\mathcal{P}$ as a full subcategory of the category $Perv$ of all perverse sheaves. We showed that a perverse sheaf N lies in $\mathcal{P}$ if and only if it admits no shifted Kummer sheaf $\mathcal{L}_\chi[1]$ as either subobject or quotient. In particular, any irreducible perverse sheaf which is not a shifted Kummer sheaf $\mathcal{L}_\chi[1]$ lies in $\mathcal{P}$.

One disadvantage of this point of view was that if

$$0 \to A \to B \to C \to 0$$

was a short exact sequence of perverse sheaves which all lay in $\mathcal{P}$, it was not the case that the sequence of their middle convolutions with an object N in $\mathcal{P}$ was necessarily exact.

Gabber and Loeser [Ga-Loe] took a different point of view. They defined a perverse sheaf N to be negligible if its Euler characteristic

$\chi(\mathbb{G}_m/\overline{k}, N)$ vanished. [The negligible N are precisely the objects $\mathcal{F}[1]$ with $\mathcal{F}$ a lisse sheaf which is a successive extension of Kummer sheaves $\mathcal{L}_\chi$ attached to characters χ of $\pi_1^{tame}(\mathbb{G}_m/\overline{k})$, cf. [**Ka-ESDE**, 8.5.2] and [**Ka-ACT**, the end of the proof of 2.5.3].] The negligible N form a thick [**Ga-Loe**, 3.6.2] subcategory Neg of the abelian category $Perv$, and they showed [**Ga-Loe**, 3.75] that the quotient category $Perv/Neg$ was an abelian category on which the two middle convolutions existed, coincided, and made $Perv/Neg$ into a neutral Tannakian category, with δ_1 as the unit object, $M \mapsto [x \mapsto 1/x]^\star DM$ as the dual, and $\chi(\mathbb{G}_m/\overline{k}, M)$ as the dimension. They further showed [**Ga-Loe**, 3.7.2] that the composition of the inclusion $\mathcal{P} \subset Perv$ followed by the passage to the quotient $Perv \to Perv/Neg$ gives an equivalence of categories

$$\mathcal{P} \cong Perv/Neg$$

under which middle convolution on $\mathcal{P}$ becomes "the" convolution on $Perv/Neg$. The upshot for $\mathcal{P}$ is while it remains a full subcategory of $Perv$, the correct structure of abelian category on it decrees that a sequence

$$0 \to A \xrightarrow{\alpha} \to B \xrightarrow{\beta} C \to 0$$

of objects of $\mathcal{P}$ is exact if α is injective, β is surjective, $\beta \circ \alpha = 0$, and $Ker(\beta)/Im(\alpha)$ is negligible. With this notion of exactness, middle convolution with a fixed N in $\mathcal{P}$ is exact, and middle convolution makes $\mathcal{P}$ into a neutral Tannakian category, cf. [**Ga-Loe**, Remarque following 3.7.7, page 535]. The notion of an irreducible object of $\mathcal{P}$ remains unchanged; it is an irreducible perverse sheaf which lies in $\mathcal{P}$, i.e., it is an irreducible perverse sheaf which is not a shifted Kummer sheaf $\mathcal{L}_\chi[1]$. Similarly for the notion of a semisimple object of $\mathcal{P}$; it is a direct sum of irreducible perverse sheaves, each of which lies in $\mathcal{P}$ (i.e., none of which is a shifted Kummer sheaf $\mathcal{L}_\chi[1]$).

Lemma 2.1. *For a semisimple object $N \in \mathcal{P}$, the compact cohomology groups $H_c^i(\mathbb{G}_m/\overline{k}, N)$ vanish for $i \neq 0$.*

Proof. We reduce immediately to the case when N is irreducible. In this case N is either a delta sheaf δ_a supported at some point $a \in \mathbb{G}_m(\overline{k})$, in which case the assertion is obvious, or it is $\mathcal{F}[1]$ for $\mathcal{F}$ an irreducible middle extension sheaf which is not a Kummer sheaf, so in particular is not geometrically constant. Then $H_c^i(\mathbb{G}_m/\overline{k}, N) = H_c^{i+1}(\mathbb{G}_m/\overline{k}, \mathcal{F})$, so we must show that $H_c^a(\mathbb{G}_m/\overline{k}, \mathcal{F})$ vanishes for $a \neq 1$. Being a middle extension, $\mathcal{F}$ has no nonzero punctual sections, so its H_c^0 vanishes. Being geometrically irreducible and not constant gives the vanishing of its H_c^2. $\square$

CHAPTER 3

Fibre Functors

At this point, we introduce the fibre functor suggested by Deligne. The proof that it is in fact a fibre functor is given in the Appendix.

Theorem 3.1. (Deligne) *Denoting by*

$$j_0 : \mathbb{G}_m/\overline{k} \subset \mathbb{A}^1/\overline{k}$$

the inclusion, the construction

$$M \mapsto H^0(\mathbb{A}^1/\overline{k}, j_{0!}M)$$

is a fibre functor on the Tannakian category $\mathcal{P}$.

For any Kummer sheaf $\mathcal{L}_\chi$ on $\mathbb{G}_m/\overline{k}$, the operation $M \mapsto M \otimes \mathcal{L}_\chi$ is an autoequivalence of $\mathcal{P}$ with itself as Tannakian category. So we get the following corollary.

Corollary 3.2. *For any Kummer sheaf $\mathcal{L}_\chi$ on $\mathbb{G}_m/\overline{k}$, the construction*

$$M \mapsto H^0(\mathbb{A}^1/\overline{k}, j_{0!}(M \otimes \mathcal{L}_\chi))$$

is a fibre functor ω_χ on the Tannakian category $\mathcal{P}$.

Let us say that a Kummer sheaf $\mathcal{L}_\chi$ on $\mathbb{G}_m/\overline{k}$ is good for the object N of $\mathcal{P}$ if, denoting by $j : \mathbb{G}_m/\overline{k} \subset \mathbb{P}^1/\overline{k}$ the inclusion, the canonical "forget supports" map is an isomorphism

$$Rj_!(N \otimes \mathcal{L}_\chi) \cong Rj_\star(N \otimes \mathcal{L}_\chi).$$

Lemma 3.3. *Given a semisimple object N of $\mathcal{P}$ and a Kummer sheaf $\mathcal{L}_\chi$ on $\mathbb{G}_m/\overline{k}$, the following conditions are equivalent.*

(1) *The Kummer sheaf $\mathcal{L}_\chi$ is good for N.*
(2) *The natural "forget supports" and "restriction" maps*

$$H^0_c(\mathbb{G}_m/\overline{k}, N \otimes \mathcal{L}_\chi) \to \omega\chi(N) := H^0(\mathbb{A}^1/\overline{k}, j_{0!}(N \otimes \mathcal{L}_\chi))$$

and

$$\omega\chi(N) := H^0(\mathbb{A}^1/\overline{k}, j_{0!}(N \otimes \mathcal{L}_\chi)) \to H^0(\mathbb{G}_m/\overline{k}, N \otimes \mathcal{L}_\chi)$$

are both isomorphisms.

(3) *The natural "forget supports" map is an isomorphism*
$$H_c^0(\mathbb{G}_m/\overline{k}, N \otimes \mathcal{L}_\chi) \cong H^0(\mathbb{G}_m/\overline{k}, N \otimes \mathcal{L}_\chi).$$

Proof. We reduce immediately to the case when N is irreducible. If N is δ_t for some point $t \in \mathbb{G}_m(\overline{k})$, then every χ is good for N, all three conditions trivially hold, and there is nothing to prove. Suppose now that N is $\mathcal{G}[1]$ for $\mathcal{G}$ an irreducible middle extension sheaf on $\mathbb{G}_m/\overline{k}$ which is not a Kummer sheaf. Replacing N by $N \otimes \mathcal{L}_\chi$, we reduce to the case when χ is the trivial character $\mathbb{1}$. Then (1) is the statement that the inertia groups $I(0)$ and $I(\infty)$ acting on $\mathcal{G}$ have neither nonzero invariants nor coinvariants, i.e., that $\mathcal{G}^{I(0)} = H^1(I(0), \mathcal{G}) = 0$ and $\mathcal{G}^{I(\infty)} = H^1(I(\infty), \mathcal{G}) = 0$. We factor j as $j_\infty \circ j_0$, where j_0 is the inclusion of $\mathbb{G}_m$ into $\mathbb{A}^1$, and j_∞ is the inclusion of $\mathbb{A}^1$ into $\mathbb{P}^1$. We have a short exact sequence of sheaves
$$0 \to j_! \mathcal{G} = j_{\infty!} j_{0!} \mathcal{G} \to j_{\infty\star} j_{0!} \mathcal{G} \to \mathcal{G}^{I(\infty)} \to 0,$$
where $\mathcal{G}^{I(\infty)}$ is viewed as a punctual sheaf supported at ∞. We view this as a short exact sequence of perverse sheaves
$$0 \to \mathcal{G}^{I(\infty)} \to j_! \mathcal{G}[1] = j_{\infty!} j_{0!} \mathcal{G}[1] \to j_{\infty\star} j_{0!} \mathcal{G}[1] \to 0.$$
Similarly, we have a short exact sequence of perverse sheaves
$$0 \to j_{\infty\star} j_{0!} \mathcal{G}[1] \to Rj_{\infty\star} j_{0!} \mathcal{G}[1] \to H^1(I(\infty), \mathcal{G}) \to 0,$$
where now $H^1(I(\infty), \mathcal{G})$ is viewed as a punctual sheaf supported at ∞. Taking their cohomology sequences on $\mathbb{P}^1$, we get short exact sequences
$$0 \to \mathcal{G}^{I(\infty)} \to H_c^0(\mathbb{G}_m/\overline{k}, \mathcal{G}[1]) \to H^0(\mathbb{P}^1/\overline{k}, j_{\infty\star} j_{0!} \mathcal{G}[1]) \to 0$$
and
$$0 \to H^0(\mathbb{P}^1/\overline{k}, j_{\infty\star} j_{0!} \mathcal{G}[1]) \to H^0(\mathbb{A}^1/\overline{k}, j_{0!} \mathcal{G}[1]) \to H^1(I(\infty), \mathcal{G}) \to 0.$$
Splicing these together, we get a four term exact sequence
$$0 \to \mathcal{G}^{I(\infty)} \to H_c^0(\mathbb{G}_m/\overline{k}, \mathcal{G}[1]) \to H^0(\mathbb{A}^1/\overline{k}, j_{0!} \mathcal{G}[1]) \to H^1(I(\infty), \mathcal{G}) \to 0.$$
A similar argument, starting with $Rj_{\infty\star} j_{0!} \mathcal{G}[1]$, gives a four term exact sequence
$$0 \to \mathcal{G}^{I(0)} \to H^0(\mathbb{A}^1/\overline{k}, j_{0!} \mathcal{G}[1]) \to H^0(\mathbb{G}_m/\overline{k}, \mathcal{G}[1]) \to H^1(I(0), \mathcal{G}) \to 0.$$
These two four term exact sequences show the equivalence of (1) and (2). It is trivial that (2) implies (3). We now show that (3) implies (1).

Suppose (3) holds. Then the composition of the two maps
$$H_c^0(\mathbb{G}_m/\overline{k}, N) \to H^0(\mathbb{A}^1/\overline{k}, j_{0!} N) \to H^0(\mathbb{G}_m/\overline{k}, N)$$
is an isomorphism. Therefore the first map is injective, and this implies that $\mathcal{G}^{I(\infty)} = 0$, which in turn implies that $H^1(I(\infty), \mathcal{G}) = 0$ (since

$\mathcal{G}^{I(\infty)}$ and $H^1(I(\infty), \mathcal{G})$ have the same dimension). And the second map is surjective, which gives the vanishing of $H^1(I(0), \mathcal{G})$, and this vanishing in turn implies the vanishing of $\mathcal{G}^{I(0)}$. □

As noted at the end of the last chapter, an irreducible object of $\mathcal{P}$ is just an irreducible perverse sheaf which lies in $\mathcal{P}$, i.e., it is an irreducible perverse sheaf which is not a Kummer sheaf $\mathcal{L}_\chi[1]$. Similarly for the notion of a semisimple object of $\mathcal{P}$; it is a direct sum of irreducible perverse sheaves, each of which lies in $\mathcal{P}$ (i.e., none of which is a Kummer sheaf $\mathcal{L}_\chi[1]$). Let us denote by $\mathcal{P}_{ss}$ the full subcategory of $\mathcal{P}$ consisting of semisimple objects. This is a subcategory stable by middle convolution (because given two semisimple objects M and N in $\mathcal{P}$, each is a completely reducible representation of the Tannakian group $G_{geom, M \oplus N}$. This group is reductive, because it has a faithful completely reducible representation, namely $M \oplus N$. Then every representation of $G_{geom, M \oplus N}$ is completely reducible, in particular the one corresponding to $M \star_{mid} N$, which is thus a semisimple object in $\mathcal{P}$. For this category $\mathcal{P}_{ss}$, its inclusion into $Perv$ is exact, and the Tannakian group $G_{geom, N}$ attached to every N in $\mathcal{P}_{ss}$ is reductive.

We end this chapter by recording two general lemmas.

Lemma 3.4. *For any perverse sheaf N on $\mathbb{G}_m/\overline{k}$, whether or not in $\mathcal{P}$, the groups $H^i(\mathbb{A}^1/\overline{k}, j_{0!}N)$ vanish for $i \neq 0$, and*

$$\dim H^0(\mathbb{A}^1/\overline{k}, j_{0!}N) = \chi(\mathbb{G}_m/\overline{k}, N) = \chi_c(\mathbb{G}_m/\overline{k}, N).$$

Proof. Every perverse sheaf on $\mathbb{G}_m/\overline{k}$ is a successive extension of finitely many geometrically irreducible ones, so we reduce to the case when N is geometrically irreducible. If N is punctual, some δ_a, the assertion is obvious. If N is $\mathcal{G}[1]$ for an irreducible middle extension sheaf, then $H^i(\mathbb{A}^1/\overline{k}, j_{0!}N) = H^{i+1}(\mathbb{A}^1/\overline{k}, j_{0!}\mathcal{G})$. The group $H^2(\mathbb{A}^1/\overline{k}, j_{0!}\mathcal{G})$ vanishes because an affine curve has cohomological dimension one, and the group $H^0(\mathbb{A}^1/\overline{k}, j_{0!}\mathcal{G})$ vanishes because $\mathcal{G}$ has no nonzero punctual sections on $\mathbb{G}_m$. Once we have this vanishing, we have $\dim H^0(\mathbb{A}^1/\overline{k}, j_{0!}N) = \chi(\mathbb{A}^1/\overline{k}, j_{0!}N)$. Then $\chi(\mathbb{A}^1/\overline{k}, j_{0!}N) = \chi_c(\mathbb{A}^1/\overline{k}, j_{0!}N)$ because $\chi = \chi_c$ on a curve (and indeed quite generally, cf. [**Lau-CC**]). Tautologically we have $\chi_c(\mathbb{A}^1/\overline{k}, j_{0!}N) = \chi_c(\mathbb{G}_m/\overline{k}, N)$, and again $\chi_c(\mathbb{G}_m/\overline{k}, N) = \chi(\mathbb{G}_m/\overline{k}, N)$. □

Lemma 3.5. *For any perverse sheaf N on $\mathbb{G}_m/\overline{k}$, whether or not in $\mathcal{P}$, the groups $H^i_c(\mathbb{G}_m/\overline{k}, N)$ vanish for $i < 0$ and for $i > 1$.*

Proof. Using the long exact cohomology sequence, we reduce immediately to the case when N is irreducible. If N is punctual, the assertion is obvious. If N is a middle extension $\mathcal{F}[1]$, then $\mathcal{F}$ has no nonzero punctual sections, i.e., $H_c^{-1}(\mathbb{G}_m/\overline{k}, \mathcal{F}[1]) = H_c^0(\mathbb{G}_m/\overline{k}, \mathcal{F}) = 0$; the groups $H_c^i(\mathbb{G}_m/\overline{k}, \mathcal{F}[1]) = H_c^{i+1}(\mathbb{G}_m/\overline{k}, \mathcal{F})$ with $i \leq -2$ or $i \geq 2$ vanish trivially. $\qquad\square$

CHAPTER 4

The Situation over a Finite Field

Let us now turn our attention to the case of a finite field k, and a groupscheme G/k which is a form of $\mathbb{G}_m$. Concretely, it is either $\mathbb{G}_m/k$ itself, or it is the nonsplit form, defined in terms of the unique quadratic extension k_2/k inside the chosen $\overline{k}$ as follows: for any k-algebra A,

$$G(A) := \{x \in A \otimes_k k_2 | \operatorname{Norm}_{A \otimes_k k_2/A}(x) = 1\}.$$

We denote by $\mathcal{P}_{arith}$ the full subcategory of the category $Perv_{arith}$ of all perverse sheaves on G/k consisting of those perverse sheaves on G/k which, pulled back to $G/\overline{k}$, lie in $\mathcal{P}$. And we denote by Neg_{arith} the full subcategory of $Perv_{arith}$ consisting of those objects which, pulled back to $G/\overline{k}$, lie in Neg. Then once again we have

$$\mathcal{P}_{arith} \cong Perv_{arith}/Neg_{arith},$$

which endows $\mathcal{P}_{arith}$ with the structure of abelian category. Thus a sequence

$$0 \to A \xrightarrow{\alpha} \to B \xrightarrow{\beta} C \to 0$$

of objects of $\mathcal{P}_{arith}$ is exact if and only if it is exact when pulled back to $G/\overline{k}$; equivalently, if and only if α is injective, β is surjective, $\beta \circ \alpha = 0$, and $Ker(\beta)/Im(\alpha)$ lies in Neg_{arith}.

An irreducible object in $\mathcal{P}_{arith}$ is an irreducible object in $Perv_{arith}$ which is not negligible, i.e., whose pullback to $G/\overline{k}$ has nonzero Euler characteristic.

Middle convolution then endows $\mathcal{P}_{arith}$ with the structure of a neutral Tannakian category, indeed a subcategory of $\mathcal{P}$ such that the inclusion $\mathcal{P}_{arith} \subset \mathcal{P}$ is exact and compatible with the tensor structure.

Recall [**BBD**, 5.3.8] that an object N in $\mathcal{P}_{arith}$ which is ι-pure of weight zero is geometrically semisimple, i.e., semisimple when pulled back to $G/\overline{k}$. Recall also that the objects N in $\mathcal{P}_{arith}$ which are ι-pure of weight zero are stable by middle convolution and form a full Tannakian subcategory $\mathcal{P}_{arith, \iota \ wt=0}$ of $\mathcal{P}_{arith}$. [Indeed, if M and N in $\mathcal{P}_{arith}$ are both ι-pure of weight zero, then $M \boxtimes N$ is ι-pure of weight zero on $\mathbb{G}_m \times \mathbb{G}_m$, hence by Deligne's main theorem [**De-Weil II**, 3.3.1] $M_{\star_!}N$ is ι-mixed of weight ≤ 0, and hence $M_{\star_{mid}}N$, as a perverse quotient of $M_{\star_!}N$, is also ι-mixed of weight ≤ 0, cf. [**BBD**, 5.3.1]. But the Verdier dual $D(M_{\star_{mid}}N)$ is the middle convolution $DM_{\star_{mid}}DN$ (because duality interchanges $R\pi_!$ and $R\pi_\star$), hence it too is ι-mixed of weight ≤ 0.]

This same stability under middle convolution holds for the objects N in $\mathcal{P}_{arith}$ which are geometrically semisimple, resp. which are arithmetically semisimple, giving full Tannakian subcategories $\mathcal{P}_{arith,gss}$, resp. $\mathcal{P}_{arith,ss}$. Thus we have full Tannakian subcategories

$$\mathcal{P}_{arith,\iota\ wt=0} \subset \mathcal{P}_{arith,gss} \subset \mathcal{P}_{arith}$$

and

$$\mathcal{P}_{arith,ss} \subset \mathcal{P}_{arith,gss} \subset \mathcal{P}_{arith}.$$

Also the objects which are both ι-pure of weight zero and arithmetically semisimple form a full Tannakian subcategory $\mathcal{P}_{arith,\iota\ wt=0,ss}$.

Pick one of the two possible isomorphisms of $G/\overline{k}$ with $\mathbb{G}_m/\overline{k}$, viewed as $\mathbb{A}^1/\overline{k} \setminus 0$. Denote by $j_0 : \mathbb{G}_m/\overline{k} \subset \mathbb{A}^1/\overline{k}$ the inclusion, and denote by ω the fibre functor on $\mathcal{P}_{arith}$ defined by

$$N \mapsto \omega(N) := H^0(\mathbb{A}^1/\overline{k}, j_{0!}N).$$

The definition of this ω depends upon the choice of one of the two $\overline{k}$-points at infinity on the complete nonsingular model, call it X, of G/k. If G/k is already $\mathbb{G}_m/k$, then these two points at infinity on X are both k-rational. But if G/k is nonsplit, these two points at infinity are only k_2-rational, and they are interchanged by $Frob_k$. So it is always the case that $Frob_{k_2}$ acts on $\omega(N)$, but $Frob_k$ may not act.

Theorem 4.1. *Suppose that N in $\mathcal{P}_{arith}$ is ι-pure of weight zero and arithmetically semisimple. Then the following six conditions are equivalent.*

(1) *For $j : G/k \subset X/k$ the inclusion of G/k into its complete non-singular model, the "forget supports" map is an isomorphism $j_!N \cong Rj_\star N$.*

(2) *The natural "forget supports" and restriction maps*

$$H^0_c(G/\overline{k}, N) \to \omega(N) \to H^0(G/\overline{k}, N)$$

are both isomorphisms.

(3) *The cohomology group $\omega(N)$ is ι-pure of weight zero for the action of $Frob_{k_2}$.*

(3bis) *The cohomology group $H^0_c(G/\overline{k}, N)$ is ι-pure of weight zero for the action of $Frob_k$ (or equivalently, for the action of $Frob_{k_2}$).*

(4) *For every object M in $<N>_{arith}$, $\omega(M)$ is ι-pure of weight zero for the action of $Frob_{k_2}$.*

(4bis) *For every object M in $<N>_{arith}$, the cohomology group $H_c^0(G/\overline{k}, M)$ is ι-pure of weight zero for the action of $Frob_k$ (or equivalently, for the action of $Frob_{k_2}$).*

(5) *For every object M in $<N>_{arith}$, the "forget supports" map is an isomorphism $j_! M \cong Rj_* M$.*

(6) *For every object M in $<N>_{arith}$, the natural "forget supports" and restriction maps*

$$H_c^0(G/\overline{k}, M) \mapsto \omega(M) \to H^0(G/\overline{k}, M)$$

are both isomorphisms.

When these equivalent conditions hold, the construction

$$M \mapsto H_c^0(G/\overline{k}, M)$$

is a fibre functor on $<N>_{arith}$, on which $Frob_k$ acts and is ι-pure of weight zero.

Proof. We first show that (1), (2), and (3) are equivalent. Since N is arithmetically semisimple, we reduce immediately to the case when N is arithmetically irreducible. If N is punctual, each of (1), (2), and (3) holds, so there is nothing to prove. So it suffices to treat the case when M is $\mathcal{G}[1]$ for an arithmetically irreducible middle extension sheaf $\mathcal{G}$ which is ι-pure of weight -1. As we saw in the proof of Lemma 3.3, we have four term exact sequences

$$0 \to \mathcal{G}^{I(\infty)} \to H_c^0(\mathbb{G}_m/\overline{k}, \mathcal{G}[1]) \to H^0(\mathbb{A}^1/\overline{k}, j_{0!}\mathcal{G}[1]) \to H^1(I(\infty), \mathcal{G}) \to 0$$

and

$$0 \to \mathcal{G}^{I(0)} \to H^0(\mathbb{A}^1/\overline{k}, j_{0!}\mathcal{G}[1]) \to H^0(\mathbb{G}_m/\overline{k}, \mathcal{G}[1]) \to H^1(I(0), \mathcal{G}) \to 0.$$

In particular, we have an injection

$$\mathcal{G}^{I(0)} \subset H^0(\mathbb{A}^1/\overline{k}, j_{0!}\mathcal{G}[1])$$

and a surjection

$$H^0(\mathbb{A}^1/\overline{k}, j_{0!}\mathcal{G}[1]) \twoheadrightarrow H^1(I(\infty), \mathcal{G}).$$

Now (1) holds for $\mathcal{G}[1]$ if and only if we have either of the following two equivalent conditions:

(a) $\mathcal{G}^{I(0)} = 0 = H^1(I(0), \mathcal{G})$ and $\mathcal{G}^{I(\infty)} = 0 = H^1(I(\infty), \mathcal{G})$.

(b) $\mathcal{G}^{I(0)} = 0$ and $H^1(I(\infty), \mathcal{G}) = 0$.

[That (a) and (b) are equivalent results from the fact that $\dim \mathcal{G}^{I(0)} = \dim H^1(I(0), \mathcal{G})$ and $\dim \mathcal{G}^{I(\infty)} = \dim H^1(I(\infty), \mathcal{G})$.] So if (1) holds, then (a) holds, and the above four term exact sequences show that (2) holds. If (2) holds, then the "forget supports" map

$$H_c^0(G/\overline{k}, N) \to H^0(G/\overline{k}, N)$$

is an isomorphism, which implies that each is ι-pure of weight zero (the source being mixed of weight ≤ 0 and the target being mixed of weight ≥ 0), and hence (3) holds. If (3) holds, we claim that (b) holds. Indeed, the four term exact sequences above give us an injection

$$\mathcal{G}^{I(0)} \subset H^0(\mathbb{A}^1/\overline{k}, j_{0!}\mathcal{G}[1])$$

and a surjection

$$H^0(\mathbb{A}^1/\overline{k}, j_{0!}\mathcal{G}[1]) \twoheadrightarrow H^1(I(\infty), \mathcal{G}).$$

But $\mathcal{G}^{I(0)}$ has ι-weight ≤ -1, so must vanish, and $H^1(I(\infty), \mathcal{G})$ has ι-weight ≥ 1, so also must vanish, exactly because $\omega(M) = H^0(\mathbb{A}^1/\overline{k}, j_{0!}\mathcal{G}[1])$ is ι-pure of weight zero.

We next show that (1) and (3bis) are equivalent. We have (1) implies (3bis) (the source is mixed of weight ≤ 0, and the target is mixed of weight ≥ 0). If (3bis) holds, we infer (1) as follows. The question is geometric, so we may reduce to the case where G is $\mathbb{G}_m$. Then on $\mathbb{P}^1$ we have the short exact sequence

$$0 \to j_!\mathcal{G} \to j_*\mathcal{G} \to \delta_0 \otimes V \bigoplus \delta_\infty \otimes W \to 0,$$

with V and W representations of $Gal(\overline{k}/k)$ which, by [**De-Weil II**, 1.81], are mixed of weight ≤ -1. Because $\mathcal{G}[1]$ has $\mathcal{P}$, $\mathcal{G}$ has no constant subsheaf, and so the group $H^0(\mathbb{P}^1, j_*\mathcal{G})$ vanishes. So the cohomology sequence gives an inclusion

$$V \oplus W \subset H^1_c(G/\overline{k}, \mathcal{G}) := H^0_c(G/\overline{k}, N).$$

As the second group is pure of weight 0, we must have $V = W = 0$. Thus we have

$$j_!\mathcal{G} \cong j_*\mathcal{G},$$

and this in turn implies that

$$j_!\mathcal{G} \cong Rj_*\mathcal{G}.$$

Each object M in $<N>_{arith}$ is itself ι-pure of weight zero and arithmetically semisimple, so applying the argument above object by object we get the equivalence of (4), (4bis), (5), and (6). Trivially (5) implies (2). Conversely, if (2) holds, then (5) holds. Indeed, it suffices to check (5) for each arithmetically irreducible object M in $<N>_{arith}$, (i.e., for any irreducible representation of the reductive group $G_{arith,N}$), but any such M is a direct factor of some multiple middle convolution of N and its dual, so its $\omega(M)$ lies in some $\omega(N)^{\otimes r} \otimes (\omega(N)^\vee)^{\otimes s}$, so is ι-pure of weight zero.

Once we have (4) and (6), the final conclusion is obvious. $\square$

Given a finite extension field E/k, and a $\overline{\mathbb{Q}}_\ell^\times$-valued character ρ of the group $G(E)$, we have the Kummer sheaf $\mathcal{L}_\rho$ on G/E, defined by pushing out the Lang torsor $1 - Frob_E : G/E \to G/E$, whose structural group is $G(E)$, by ρ. For L/E a finite extension, and $\rho_L := \rho \circ Norm_{L/E}$, the Kummer sheaf $\mathcal{L}_{\rho_L}$ on G/L is the pullback of $\mathcal{L}_\rho$ on G/E. Pulled back to $G/\overline{k}$, the Kummer sheaves $\mathcal{L}_\rho$ are just the characters of finite order of the tame fundamental group $\pi_i^{tame}(G/\overline{k})$; here we view this group as the inverse limit, over finite extensions E/k, of the groups $G(E)$, with transition maps provided by the norm.

We say that a character ρ of some $G(E)$ is good for an object N in $\mathcal{P}_{arith}$ if this becomes true after extension of scalars to $G/\overline{k}$, i.e., if the "forget supports" map is an isomorphism $j_!(N \otimes \mathcal{L}_\rho) \cong Rj_*(N \otimes \mathcal{L}_\rho)$. As an immediate corollary of the previous theorem, applied to $N \otimes \mathcal{L}_\rho$ on G/E, we get the following.

Corollary 4.2. *Suppose N in $\mathcal{P}_{arith}$ is ι-pure of weight zero and arithmetically semisimple. If a character ρ of some $G(E)$ is good for N, then for every M in $<N>_{arith}$, $H_c^0(G/\overline{k}, M \otimes \mathcal{L}_\rho)$ is ι-pure of weight zero for $Frob_E$, and the construction*

$$M \mapsto H_c^0(G/\overline{k}, M \otimes \mathcal{L}_\rho)$$

is a fibre functor on $<N>_{arith}$, on which $Frob_E$ acts and is ι-pure of weight zero.

CHAPTER 5

Frobenius Conjugacy Classes

Let G/k be a form of $\mathbb{G}_m$, and N in $\mathcal{P}_{arith}$ an object which is ι-pure of weight zero and arithmetically semisimple. If G/k is $\mathbb{G}_m/k$, then we have the fibre functor on $<N>_{arith}$ given by

$$M \mapsto \omega(N) := H^0(\mathbb{A}^1/\overline{k}, j_{0!}M),$$

on which $Frob_k$ operates. And for each finite extension field E/k and each character χ of $G(E)$, we have the fibre functor ω_χ on $<N>_{arith}$ given by

$$\omega_\chi(M) := \omega(M \otimes \mathcal{L}_\chi),$$

on which $Frob_E$ operates. This action of $Frob_E$ on ω_χ gives us an element in the Tannakian group G_{arith,N,ω_χ} for $<N>_{arith}$, and so a conjugacy class $Frob_{E,\chi}$ in the reference Tannakian group $G_{arith,N} := G_{arith,N,\omega}$. By the definition of this conjugacy class, we have an identity of characteristic polynomials

$$\det(1 - TFrob_{E,\chi}|\omega(N)) = \det(1 - TFrob_E|\omega_\chi(N)).$$

When G/k is the nonsplit form, then the definition of ω depends upon a choice, and so in general only $Frob_{k_2}$ acts on it. For a finite extension field E/k and a character χ of $G(E)$, we have the fibre functor ω_χ on $<N>_{arith}$ given by

$$\omega_\chi(M) := \omega(M \otimes \mathcal{L}_\chi),$$

but we have $Frob_E$ acting on it only if either $deg(E/k)$ is even (in which case G/E is $\mathbb{G}_m/E$) **or** if χ is good for N, in which case ω_χ is the fibre functor

$$M \mapsto H_c^0(G/\overline{k}, M \otimes \mathcal{L}_\chi),$$

on which $Frob_E$ acts. So we get conjugacy classes $Frob_{E,\chi}$ in the reference Tannakian group $G_{arith,N}$ when either χ is good for N or when E/k has even degree. And again here we have an identity of characteristic polynomials

$$\det(1 - TFrob_{E,\chi}|\omega(N)) = \det(1 - TFrob_E|\omega_\chi(N)).$$

In either the split or nonsplit case, when χ is good for N, the conjugacy class $Frob_{E,\chi}$ has unitary eigenvalues in every representation of the reductive group $G_{arith,N}$. Now fix a maximal compact subgroup K of the complex reductive group $G_{arith,N}(\mathbb{C})$. As explained in the introduction, the semisimple part (in the sense of Jordan decomposition) of $Frob_{E,\chi}$ gives rise to a well-defined conjugacy class $\theta_{E,\chi}$ in K.

As we will see later when we try to compute examples, the Frobenius conjugacy classes $Frob_{E,\chi}$ in $G_{arith,N}$ attached to χ's which are **not good** for N will also play a key role, providing a substitute for local monodromy. But for the time being we focus on the classes $\theta_{E,\chi}$ in K.

CHAPTER 6

Group-Theoretic Facts about G_{geom} and G_{arith}

Theorem 6.1. *Suppose N in $\mathcal{P}_{arith}$ is geometrically semisimple. Then $G_{geom,N}$ is a normal subgroup of $G_{arith,N}$.*

Proof. Because N is geometrically semisimple, the group $G_{geom,N}$ is reductive, so it is the fixer of its invariants in all finite dimensional representations of the ambient $G_{arith,N}$. By noetherianity, there is a finite list of representations of $G_{arith,N}$ such that $G_{geom,N}$ is the fixer of its invariants in these representations. Taking the direct sum of these representations, we get a single representation of $G_{arith,N}$ such that $G_{geom,N}$ is the fixer of its invariants in that single representation. This representation corresponds to an object M in $<N>_{arith}$, and a $G_{geom,N}$-invariant in that representation corresponds to a δ_1 sitting inside M_{geom}. So the entire space of $G_{geom,N}$-invariants corresponds to the subobject $Hom_{geom}(\delta_1, M) \otimes \delta_1$ of M. [This is an arithmetic subobject, of the form (a $Gal(\overline{k}/k)$-representation)$\otimes \delta_1$.] Thus the space of $G_{geom,N}$-invariants is $G_{arith,N}$-stable (because it corresponds to an arithmetic subobject of M). But the fixer of any $G_{arith,N}$-stable subspace in any representation of $G_{arith,N}$ is a normal subgroup of $G_{arith,N}$. $\square$

Theorem 6.2. *Suppose that N in $\mathcal{P}_{arith}$ is arithmetically semisimple and pure of weight zero (i.e., ι-pure of weight zero for every ι). If $G_{arith,N}$ is finite, then N is punctual. Indeed, if every Frobenius conjugacy class $Frob_{E,\chi}$ in $G_{arith,N}$ is quasiunipotent (:= all eigenvalues are roots of unity), then N is punctual.*

Proof. We argue by contradiction. If N is not punctual, it has a non-punctual irreducible constituent. The G_{arith} of this constituent is a quotient of $G_{arith,N}$ which inherits the quasiunipotence property. So we reduce to the case when N is $\mathcal{G}[1]$ for an (arithmetically irreducible, but we will not use this) middle extension sheaf $\mathcal{G}$ which is pure of weight -1. As every $Frob_{E,\chi}$ is quasiunipotent, in particular it has unitary eigenvalues, so every fibre functor ω_χ is pure of weight zero, and hence every χ is good for N, and its fibre functor is just

$$\omega_\chi(M) \cong H^0_c(G/\overline{k}, M \otimes \mathcal{L}_\chi).$$

On some dense open set $U \subset G$, $\mathcal{G}|U$ is a lisse sheaf on U, pure of weight -1 and of some rank $r \geq 1$. For any large enough finite field extension E/k, $U(E)$ is nonempty. Pick such an E/k of even degree, and a point $a \in U(E)$. For each character χ of $G(E) \cong E^\times$, we have

$$\mathrm{Trace}(Frob_{E,\chi}|\omega(N)) = \mathrm{Trace}(Frob_E|H_c^0(G/\overline{k}, \mathcal{G}[1] \otimes \mathcal{L}_\chi)$$

$$= - \sum_{t \in G(E)} \chi(t)\mathrm{Trace}(Frob_{E,t}|\mathcal{G}).$$

By multiplicative Fourier inversion, we get

$$\#G(E)\mathrm{Trace}(Frob_{E,a}|\mathcal{G}) = - \sum_{\chi \text{ char of } G(E)} \overline{\chi}(a)\mathrm{Trace}(Frob_{E,\chi}|\omega(N)).$$

And for the finite extension E_n/E of degree n, we get

$$\#G(E_n)\mathrm{Trace}((Frob_{E,a})^n|\mathcal{G}) = - \sum_{\chi \text{ char of } G(E_n)} \overline{\chi}(a)\mathrm{Trace}(Frob_{E_n,\chi}|\omega(N)).$$

For each n, $\#G(E_n) = (\#E)^n - 1$. So in terms of the r eigenvalues $\alpha_1, ..., \alpha_r$ of $Frob_{E,a}|\mathcal{G}$, we find that for every $n \geq 1$, the quantity

$$((\#E)^n - 1)(\sum_i (\alpha_i)^n)$$

is a cyclotomic integer (because each $Frob_{E_n,\chi}|\omega(N)$ is quasiunipotent).

From the relation of the first r Newton symmetric functions to the standard ones, we see that the characteristic polynomial of $Frob_{E,a}|\mathcal{G}$ has coefficients in some cyclotomic field, call it L. Hence all Newton symmetric functions of the α_i lie in L. Now consider the rational function of one variable T,

$$\sum_i 1/(1 - \alpha_i T) - \sum_i 1/(1 - (\#E)\alpha_i T) = \sum_{n \geq 1} A_n T^n,$$

where the coefficients $A_n = -((\#E)^n - 1)(\sum_i (\alpha_i)^n)$ are, on the one hand, cyclotomic integers, and on the other hand lie in the fixed cyclotomic field L. Therefore the A_n lie in $\mathcal{O}_L$, the ring of integers in L. Therefore the power series around 0 for this rational function converges λ-adically in $|T|_\lambda < 1$, for every nonarchimedean place λ of any algebraic closure of L.

By purity, each α_i has complex absolute value $1/\sqrt{\#E}$, and hence for all i, j we have $\alpha_i \neq (\#E)\alpha_j$. So there is no cancellation in the expression

$$\sum_i 1/(1 - \alpha_i T) - \sum_i 1/(1 - (\#E)\alpha_i T)$$

of our rational function. Hence for each i, $1/\alpha_i$ is a pole. Therefore we have

$$|1/\alpha_i|_\lambda \geq 1$$

for each nonarchimedean λ, which is to say

$$|\alpha_i|_\lambda \leq 1$$

for each nonarchimedean λ. Thus each α_i is an algebraic integer (in some finite extension of L). But **every** archimedean absolute value of α_i is $1/\sqrt{\#E} < 1$. This violates the product formula, and so achieves the desired contradiction. □

Corollary 6.3. *Suppose that N in $\mathcal{P}_{arith}$ is arithmetically semisimple and pure of weight zero (i.e., ι-pure of weight zero for every ι). Suppose further that its Tannakian determinant $M :=$ "$\det$"(N) has $G_{geom,M}$ finite. If the group $G_{arith,N}/(G_{arith,N} \cap$ scalars$)$ is finite, then N is punctual. Indeed, if every Frobenius conjugacy class $Frob_{E,\chi}$ in $G_{arith,N}$ has a power which has all equal eigenvalues, then N is punctual.*

Proof. $M :=$ "$\det$"(N) is a one-dimensional object, i.e., an object of $<N>_{arith}$ with $\chi(\mathbb{G}_m/\overline{k}, M) = 1$. One knows [**Ka-ESDE**, 8.5.3] that the only such objects on $G/\overline{k} \cong \mathbb{G}_m/\overline{k}$ are either punctual objects δ_a for some $a \in G(\overline{k})$, or are multiplicative translates of hypergeometric sheaves $\mathcal{H}(\psi, \chi_1, ..., \chi_n; \rho_1, ..., \rho_m)[1]$, where $\text{Max}(n, m) \geq 1$ and where, for all i, j, we have $\chi_i \neq \rho_j$. But such a hypergeometric sheaf has its G_{geom} equal to the group $GL(1)$. Indeed, by [**Ka-ESDE**, 8.2.3] and [**Ka-ESDE**, 8.4.2 (5)], for any integer $r \geq 1$, the r-fold middle convolution of $\mathcal{H}(\psi, \chi_1, ..., \chi_n; \rho_1, ..., \rho_m)[1]$ with itself is the hypergeometric sheaf of type (rn, rm) given by

$$\mathcal{H}(\psi, \text{each } \chi_i \text{ repeated r times}; \text{each } \rho_j \text{ repeated r times})[1].$$

Thus no Tannakian tensor power of $\mathcal{H}(\psi, \chi_1, ..., \chi_n; \rho_1, ..., \rho_m)[1]$ is trivial, so its G_{geom} must be the entire group $GL(1)$.

Therefore M, having finite G_{geom}, is geometrically some δ_a, for some $a \in G(\overline{k})$. But M lies in $<N>_{arith}$, so we must have $a \in G(k)$, and M is arithmetically $\alpha^{deg} \otimes \delta_a$, for some unitary α.

The statement to be proven, that N is punctual, is a geometric one. And our hypotheses remain valid if we pass from k to any finite extension field E/k. Doing so, we may reduce to the case when G/k is split. Let us denote by $d =$ "$\dim$"$N := \chi(\mathbb{G}_m/\overline{k}, N)$ the "dimension" of N. Choose a d'th root β of $1/\alpha$, and a d'th root b of $1/a$ in some finite extension field of k. Again extending scalars, we may reduce to the case when $b \in k^\times$. Now consider the object $N \star_{mid} (\beta^{deg} \otimes \delta_b)$, on G/k. This object satisfies all our hypotheses, but now its determinant is arithmetically trivial. So if every $Frob_{E,\chi}$ has a power with equal eigenvalues, those equal eigenvalues must be d'th roots of unity, since the determinant is 1. Hence every $Frob_{E,\chi}$ is quasiunipotent, and we conclude by the previous result that $N \star_{mid} \beta^{deg} \otimes \delta_b$ is punctual. Hence N itself is punctual.   □

Theorem 6.4. *Suppose that N in $\mathcal{P}_{arith}$ is arithmetically semisimple and pure of weight zero (i.e., ι-pure of weight zero for every ι). If $G_{geom,N}$ is finite, then N is punctual.*

Proof. We argue by contradiction. If N is not punctual, it has some arithmetically irreducible constituent M which is not punctual. Then $G_{geom,M}$ is finite, being a quotient of $G_{geom,N}$. So we are reduced to the case when M is arithmetically irreducible, of the form $\mathcal{G}[1]$ for an arithmetically irreducible middle extension sheaf $\mathcal{G}$.

We wish to reduce further to the case in which $\mathcal{G}$ is geometrically irreducible. Think of $\mathcal{G}$ as the extension by direct image of an arithmetically irreducible lisse sheaf $\mathcal{F}$ on a dense open set $U \subset G$. If $\mathcal{F}|\pi_1^{geom}(U)$ is $\bigoplus n_i \mathcal{F}_i$, with the $\mathcal{F}_i$ inequivalent irreducible representations of $\pi_1^{geom}(U)$, then $\pi_1^{arith}(U)$ must transitively permute the isotypical components $n_i \mathcal{F}_i$. Passing to a finite extension field E/k, we reduce to the case when each isotypical component $n_i \mathcal{F}_i$ is a $\pi_1^{arith}(U)$-representation.

Passing to one isotypical component, we reduce to the case when N is geometrically $\mathcal{G}[1]$ for a middle extension sheaf $\mathcal{G}$, itself the extension by direct image of a lisse sheaf $\mathcal{F}$ on U/k such that $\mathcal{F} = n\mathcal{F}_1$ is geometrically isotypical. Because we have extended scalars, $\mathcal{F}$ may not be arithmetically irreducible, but each of its arithmetically irreducible constituents is itself geometrically isotypical, of the form $n\mathcal{F}_1$ for some possibly lower value of n. So it suffices to treat the case in which $\mathcal{F}$ is both arithmetically irreducible and geometrically isotypical (i.e., geometrically $n\mathcal{F}_1$ for some geometrically irreducible $\mathcal{F}_1$ and some $n \geq 1$). We claim that in fact $n = 1$, i.e., that $\mathcal{F}$ is geometrically irreducible. To see this, we argue as follows. Consider the dense subgroup Γ of $\pi_1^{arith}(U)$ is given by the semidirect product $\Gamma := \pi_1^{geom}(U) \rtimes F^{\mathbb{Z}}$, where we take for F an element of degree one in $\pi_1^{arith}(U)$. Because Γ is dense in $\pi_1^{arith}(U)$, $\mathcal{F}$ is Γ-irreducible. So the isomorphism class of $\mathcal{F}_1$ must be invariant by Γ. In other words, $\mathcal{F}_1$ is a representation of $\pi_1^{geom}(U)$ whose isomorphism class is invariant by F, and hence $\mathcal{F}_1$ admits a structure of Γ-representation. As representations of Γ, we have

$$\mathcal{F} \cong \mathcal{F}_1 \otimes Hom_{geom}(\mathcal{F}_1, \mathcal{F}).$$

Here $Hom_{geom}(\mathcal{F}_1, \mathcal{F})$ is an $F^{\mathbb{Z}}$-representation. It must be irreducible, because $\mathcal{F}$ is Γ-irreducible. But $n = \dim Hom_{geom}(\mathcal{F}_1, \mathcal{F})$, hence $n = 1$ (because every irreducible representation of the abelian group $F^{\mathbb{Z}}$ has dimension one). Therefore $\mathcal{F}$ itself is geometrically irreducible, and so $\mathcal{G}[1]$ is geometrically irreducible.

So our situation is that we have an object $N = \mathcal{G}[1]$ which is both arithmetically and geometrically irreducible, pure of weight zero, and whose $G_{geom,N}$ is finite. Therefore its Tannakian determinant $M := \det(N)$ has $G_{geom,M}$ finite. Now $G_{geom,N}$ acts irreducibly in the representation corresponding to N. But $G_{geom,N}$ is normal in $G_{arith,N}$, so every element of $G_{arith,N}$ normalizes the finite irreducible group $G_{geom,N}$. But $Aut(G_{geom,N})$ is certainly finite, so a (fixed) power of every element in $G_{arith,N}$ commutes with the irreducible group $G_{geom,N}$, so is scalar. In particular, each $Frob_{E,\chi}$ has a power which is scalar. The desired contradiction then results from the previous Corollary 6.3. $\square$

Theorem 6.5. *Suppose that N in $\mathcal{P}_{arith}$ is arithmetically semisimple and pure of weight zero (i.e., ι-pure of weight zero for every ι). Then the group $G_{geom,N}/G^0_{geom,N}$ of connected components of $G_{geom,N}$ is cyclic of some prime to p order n. Its order n is the order of the group*

$$\{\zeta \in \overline{k}^{\times} | \delta_{\zeta} \in\, < N >_{geom}\}.$$

The irreducible punctual objects in $< N >_{geom}$ are precisely the objects δ_{ζ} with $\zeta \in \mu_n(\overline{k})$.

Proof. Since $G_{geom,N}$ is normal in $G_{arith,N}$, $G^0_{geom,N}$ is also normal in $G_{arith,N}$. So we may take a faithful representation of the quotient group $G_{arith,N}/G^0_{geom,N}$; this is an object M in $\mathcal{P}_{arith}$ which is arithmetically semisimple and pure of weight zero, and whose $G_{geom,M}$ is given by $G_{geom,M} = G_{geom,N}/G^0_{geom,N}$. Hence M is punctual, so geometrically a direct sum of finitely many objects δ_{a_i} for various a_i in $\overline{k}^{\times}$. But each a_i lies in a finite field, hence the group generated by the a_i is a finite subgroup of $\overline{k}^{\times}$, so it is the group $\mu_n(\overline{k})$ for some prime to p integer n. So the objects in $<M>_{geom}$ are just the direct sums of finitely many objects δ_{ζ_i}, for various $\zeta_i \in \mu_n(\overline{k})$, i.e., they are the representations of $\mathbb{Z}/n\mathbb{Z}$. But this same M is a faithful representation of the smaller group $G_{geom,N}/G^0_{geom,N}$. Hence $G_{geom,N}/G^0_{geom,N} = \mathbb{Z}/n\mathbb{Z}$, and the irreducible punctual objects in $< N >_{geom}$ are precisely the objects δ_{ζ} with $\zeta \in \mu_n(\overline{k})$. $\square$

Corollary 6.6. *Suppose that N in $\mathcal{P}_{arith}$ is arithmetically semisimple and pure of weight zero (i.e., ι-pure of weight zero for every ι). Then $G^0_{geom,N}$, viewed inside $G_{geom,N}$, is the fixer of all punctual objects in*

$<N>_{geom}$. *In particular, $G_{geom,N}$ is connected if and only if the only irreducible punctual object in $<N>_{geom}$ is δ_1.*

Proof. Because $G^0_{geom,N}$ is normal in the reductive group $G_{geom,N}$, it is itself reductive, so is the fixer of its invariants in $<N>_{geom}$. We must show that its invariants there are precisely the punctual objects in $<N>_{geom}$.

Any punctual object in $<N>_{geom}$ is a sum of objects δ_ζ, with $\zeta \in \overline{k}^\times$ (necessarily) a root of unity, so $G^0_{geom,N}$ acts trivially on any punctual object in $<N>_{geom}$. Conversely, suppose $L \in <N>_{geom}$ is an irreducible object on which $G^0_{geom,N}$ acts trivially. We must show that L is punctual. In any case, L, like any irreducible object of $<N>_{geom}$, is geometrically a direct factor of some object of $<N>_{arith}$, indeed of some direct sum of multiple middle convolutions of N and its Tannakian dual $N^\vee$. So L is geometrically a direct factor of some arithmetically irreducible object M of $<N>_{arith}$. If M is punctual, then L is punctual. If M is not punctual, we get a contradiction as follows: if M is not punctual, then it is $j_*\mathcal{F}[1]$ for some arithmetically irreducible lisse sheaf on some dense open set $j : U \subset G$. Exactly as in the proof of Theorem 6.4, after a finite extension of the ground field, we reduce to the case when M is geometrically irreducible, hence geometrically isomorphic to L. But M is not punctual, so not fixed by $G^0_{geom,N}$, contradiction. $\square$

CHAPTER 7

The Main Theorem

Lemma 7.1. *Let G/k be a form of $\mathbb{G}_m$, and N in $\mathcal{P}_{arith}$ ι-pure of weight zero and arithmetically semisimple. The quotient group $G_{arith,N}/G_{geom,N}$ is a group of multiplicative type, in which a Zariski dense subgroup is generated by the image of any single Frobenius conjugacy class $Frob_{k,\chi}$. If the quotient is finite, say of order n, then it is canonically $\mathbb{Z}/n\mathbb{Z}$, and the image in this quotient of any Frobenius conjugacy class $Frob_{E,\chi}$ is $deg(E/k)$ mod n.*

Proof. Representations of the quotient $G_{arith,N}/G_{geom,N}$ are objects in $<N>_{arith}$ which are geometrically trivial, i.e., those objects $V \otimes \delta_1$, for V some completely reducible representation of $Gal(\overline{k}/k)$, which lie in $<N>_{arith}$. Such an object is a finite direct sum of one-dimensional objects $\alpha_i^{deg} \otimes \delta_1$, for unitary scalars α_i. Take such an object which is a faithful representation Λ of the quotient $G_{arith,N}/G_{geom,N}$, say $V = \oplus_{i=1}^d \alpha_i^{deg} \otimes \delta_1$. Then the quotient $G_{arith,N}/G_{geom,N}$ is the Zariski closure, in $GL(1)^d$ of the subgroup generated by $\Lambda(Frob_k) = (\alpha_1, ..., \alpha_d)$. If this quotient is finite of order n, it is cyclic. Concretely, this means that the α_i are each n'th roots of unity, and that they generate the group $\mu_n(\overline{\mathbb{Q}_\ell})$. The image of a Frobenius conjugacy class $Frob_{E,\chi}$ in this representation is just $(\alpha_1^{deg(E/k)}, ..., \alpha_d^{deg(E/k)})$, which is $\Lambda(Frob_k)^{deg(E/k)}$. $\square$

Let us now consider an object N in $\mathcal{P}_{arith}$ which is ι-pure of weight zero and arithmetically semisimple, and such that the quotient group $G_{arith,N}/G_{geom,N}$ is $\mathbb{Z}/n\mathbb{Z}$. Choose a maximal compact subgroup K_{geom} of the reductive Lie group $G_{geom,N}(\mathbb{C})$. Because K_{geom} is a compact subgroup of $G_{arith,N}(\mathbb{C})$, we may choose a maximal compact subgroup K_{arith} of $G_{arith,N}$ such that $K_{geom} \subset K_{arith}$. Notice that K_{geom} is the intersection $G_{geom,N} \cap K_{arith}$; indeed, this intersection is compact, lies in $G_{geom,N}$, and contains K_{geom}, so by maximality of K_{geom} it must be K_{geom}. Because K_{arith} is Zariski dense in $G_{arith,N}$, it maps onto the finite quotient $\mathbb{Z}/n\mathbb{Z}$, and the kernel is K_{geom}.

So our situation is that K_{geom} is a normal subgroup of K_{arith}, with quotient $K_{arith}/K_{geom} = \mathbb{Z}/n\mathbb{Z}$. For each integer d mod n, we denote by $K_{arith,d} \subset K_{arith}$ the inverse image of d mod n. In terms of any element $\gamma_d \in K_{arith}$ of degree d, $K_{arith,d}$ is the coset $\gamma_d K_{geom}$. Because $\mathbb{Z}/n\mathbb{Z}$ is abelian, the space $K_{arith}^{\#}$ of conjugacy classes in K_{arith} maps onto $\mathbb{Z}/n\mathbb{Z}$. For each integer d mod n, we denote by $K_{arith,d}^{\#} \subset K_{arith}^{\#}$ the inverse image of d mod n. Concretely, $K_{arith,d}^{\#}$ is the quotient set of $K_{arith,d}$ by the conjugation action of the ambient group K_{arith}. Denote by μ the Haar measure dk on K_{arith} of total mass n, i.e., the Haar measure which gives K_{geom} total mass one, and denote by $\mu^{\#}$ its direct image on $K_{arith}^{\#}$. For each integer d mod n, we denote by $\mu_d^{\#}$ the restriction of $\mu^{\#}$ to $K_{arith,d}^{\#}$. We denote by $i_\star \mu_d^{\#}$ the extension by zero of $\mu_d^{\#}$ to $K_{arith}^{\#}$.

Theorem 7.2. *Suppose N in $\mathcal{P}_{arith}$ ι-pure of weight zero and arithmetically semisimple, such that the quotient group $G_{arith,N}/G_{geom,N}$ is $\mathbb{Z}/n\mathbb{Z}$. Fix an integer d mod n. Then as E/k runs over larger and larger extension fields whose degree is d mod n, the conjugacy classes $\{\theta_{E,\rho}\}_{good\ \rho}$ become equidistributed in the space $K_{arith,d}^{\#}$ for the measure $\mu_d^{\#}$ of total mass one. Equivalently, as E/k runs over larger and larger extension fields whose degree is d mod n, the conjugacy classes $\{\theta_{E,\rho}\}_{good\ \rho}$ become equidistributed in the space $K_{arith}^{\#}$ for the measure $i_\star \mu_d^{\#}$ of total mass one.*

Proof. We must show that for any continuous central function f on K_{arith} and for any integer d mod n, we can compute $\int_{K_{arith,d}} f(k)dk$ as the limit, as E/k runs over larger and larger extension fields whose degree is d mod n, large enough that the set $Good(E, N)$ is nonempty, of the sums

$$1/\#Good(E, N) \sum_{\rho \in Good(E,N)} f(\theta_{E,\rho}).$$

[Remember that we are using the Haar measure dk on K_{arith} of total mass n, so that each $K_{arith,d}$ has measure one.] By the Peter-Weyl theorem, any such f is a uniform limit of finite linear combinations of traces of irreducible representations of K_{arith}. So it suffices to check when f is the trace of an irreducible representation Λ_K of K_{arith}. If Λ_K is the trivial representation, both the integral and each of the sums is identically 1.

So it remains to show that for any irreducible nontrivial representation Λ_K of K_{arith}, and for any integer d mod n, we can compute $\int_{K_{arith,d}} \text{Trace}(\Lambda_K(k))dk$ as the limit, as E/k runs over larger and larger extension fields whose degree is d mod n, large enough that the set

$Good(E, N)$ is nonempty, of the sums

$$1/\#Good(E, N) \sum_{\rho \in Good(E,N)} \text{Trace}(\Lambda_K(\theta_{E,\rho})).$$

The representation Λ_K is the restriction to K_{arith} of a unique irreducible nontrivial representation Λ of $G_{arith,N}$. Denote by M the arithmetically irreducible nontrivial object corresponding to Λ.

Because $G_{geom,N}$ is a normal subgroup of $G_{arith,N}$, the space $\Lambda^{G_{geom,N}}$ of $G_{geom,N}$-invariants in Λ is a subrepresentation, so by $G_{arith,N}$-irreducibility it is either Λ or 0. We now treat the two cases separately.

Suppose first that $G_{geom,N}$ acts trivially. Then Λ is a nontrivial irreducible representation of $\mathbb{Z}/n\mathbb{Z}$, so it is $\alpha^{deg} \otimes \delta_1$ for some n'th root of unity $\alpha \neq 1$. Then

$$\int_{K_{arith,d}} \text{Trace}(\Lambda(k))dk = \alpha^d.$$

And for any E/k whose degree is $d \bmod n$ and large enough that the set $Good(E, N)$ is nonempty, we have an identity

$$1/\#Good(E, N) \sum_{\rho \in Good(E,N)} \text{Trace}(\Lambda(\theta_{E,\rho})) = \alpha^d,$$

indeed each individual summand $\text{Trace}(\Lambda(\theta_{E,\rho})) = \alpha^d$.

Next suppose that $\Lambda^{G_{geom,N}} = 0$. Then for any integer $d \bmod n$, we claim that

$$\int_{K_{arith,d}} \text{Trace}(\Lambda(k))dk = 0.$$

Indeed, if we fix an element $\gamma \in K_{arith,d}$, then $K_{arith,d} = \gamma K_{geom}$, so this integral is

$$\int_{K_{geom}} \text{Trace}(\Lambda(\gamma k))dk = \int_{K_{geom}} \text{Trace}(\Lambda(k\gamma))dk.$$

This last integral is, in turn, the trace of the integral operator on the representation space V of Λ given by

$$v \mapsto \int_{K_{geom}} \Lambda(k)(\Lambda(\gamma)(v))dk.$$

But the integral operator

$$w \mapsto \int_{K_{geom}} \Lambda(k)(w)dk.$$

is just the projection onto the space $V^{G_{geom,N}} = 0$ of $G_{geom,N}$-invariants in Λ. So this operator vanishes, and so also its trace vanishes.

We will now show that as E/k runs over larger and larger extensions of any degree, we have

$$(1/\#Good(E, N)) \sum_{\rho \in Good(E,N)} \text{Trace}(\Lambda(\theta_{E,\rho})) = O(1/\sqrt{\#E}).$$

For good ρ, the term $\text{Trace}(\Lambda(\theta_{E,\rho}))$ is

$$\text{Trace}(Frob_E | H_c^0(G/\overline{k}, M \otimes \mathcal{L}_\rho)).$$

For any ρ, the cohomology groups $H_c^i(G/\overline{k}, M \otimes \mathcal{L}_\rho)$ vanish for $i \neq 0$, cf. Lemma 2.1, so the Lefschetz Trace formula [**Gr-Rat**] gives

$$\text{Trace}(Frob_E | H_c^0(G/\overline{k}, M \otimes \mathcal{L}_\rho)) = \sum_{s \in G(E)} \rho(s) \text{Trace}(Frob_{E,s} | M).$$

By the main theorem [**De-Weil II**, 3.3.1] of Deligne's Weil II, $H_c^0(G/\overline{k}, M \otimes \mathcal{L}_\rho)$ is ι-mixed of weight ≤ 0, so we have the estimate

$$|\text{Trace}(Frob_E | H_c^0(G/\overline{k}, M \otimes \mathcal{L}_\rho))| \leq \text{``dim''}(M).$$

So the sum

$$(1/\#Good(E, N)) \sum_{\rho \in Good(E,N)} \text{Trace}(\Lambda(\theta_{E,\rho}))$$

is within $O(1/\#E)$ of the sum

$$(1/\#G(E)) \sum_{\rho \in G(E)^\vee} \text{Trace}(Frob_E | H_c^0(G/\overline{k}, M \otimes \mathcal{L}_\rho)).$$

More precisely, we have

$$(1/\#Good) \sum_{Good} = (1/\#All) \sum_{All}$$

$$+(1/\#Good - 1/\#All) \sum_{All}$$

$$-(1/\#Good) \sum_{Bad}.$$

Expanding each summand by the Lefschetz Trace formula [**Gr-Rat**], we see that the first sum $(1/\#All) \sum_{All}$ is

$$\text{Trace}(Frob_{E,1} | M).$$

The second term is

$$(1/\#Good - 1/\#All) \sum_{All} = (\#Bad/\#Good) \text{Trace}(Frob_{E,1} | M).$$

The final term is bounded by

$$|-(1/\#Good)\sum_{Bad}| \le (\#Bad/\#Good)\text{``dim''}(M).$$

Here we must distinguish two subcases. It may be that M, the object corresponding to Λ, is punctual. In that case, as M is not geometrically trivial, it must be $\alpha^{deg}\otimes\delta_Z$ for Z a closed point of G which is not the closed point 1. Then $\text{Trace}(Frob_{E,1}|M) = 0$. If M is not punctual, then it is $\mathcal{F}[1]$ for an (arithmetically irreducible, but we will not use this) middle extension sheaf which is ι-pure of weight -1. Then $\text{Trace}(Frob_{E,1}|M) = -\text{Trace}(Frob_{E,1}|\mathcal{F})$ has $|-\text{Trace}(Frob_{E,1}|\mathcal{F})| \le rk(\mathcal{F}_1)/\sqrt{\#E} \le gen.rk(\mathcal{F}_1)/\sqrt{\#E}$, by [**De-Weil II**, 1.8.1], where we have written $gen.rk(\mathcal{F})$ for the rank of the restriction of $\mathcal{F}$ to a dense open set where it is lisse. $\qquad\square$

In the special case when $G_{geom,N} = G_{arith,N}$, we get the following theorem and its corollary.

Theorem 7.3. *Let G/k be a form of $\mathbb{G}_m$, and N in $\mathcal{P}_{arith}$ ι-pure of weight zero and arithmetically semisimple. Choose a maximal compact subgroup K of the reductive Lie group $G_{arith,N}(\mathbb{C})$. Suppose that we have an* **equality** *of groups*

$$G_{geom,N} = G_{arith,N}.$$

Then as E/k runs over larger and larger finite extension fields, the conjugacy classes $\{\theta_{E,\rho}\}_{good\ \rho}$ become equidistributed in the space $K^\#$ of conjugacy classes in K, for the induced Haar measure of total mass one.

Corollary 7.4. *As E/k runs over larger and larger finite extension fields, the sums $\{S(N, E, \rho)\}_{good\ \rho}$ defined by*

$$S(N, E, \rho) := \sum_{t\in G(E)} \rho(t)\text{Trace}(Frob_{E,t}|N),$$

become equidistributed in $\mathbb{C}$ for the probability measure which is the direct image by the Trace map

$$\text{Trace}: K \to \mathbb{C}$$

of Haar measure of total mass one on K.

Remark 7.5. Suppose we are in the situation of Theorem 7.3. Let us denote by $Bad(N)$ the finite set of characters of $\pi_1^{tame}(\mathbb{G}_m \otimes_k \overline{k})$ which are bad for N. For any object M in $\mathcal{P}_{geom}$, let us denote by $gen.rk(M)$ the integer defined as follows: on an open dense set U of $G_{\overline{k}}$, M is $\mathcal{F}[1]$ for some lisse sheaf $\mathcal{F}$ on U, and we define the rank of $\mathcal{F}$ to be

$gen.rk(M)$. Suppose E is large enough that $\#Bad(N) \le \sqrt{\#E} - 1$. Then for an irreducible nontrivial representation Λ of $G_{arith,N}$, corresponding to an object $M \in \mathcal{P}_{geom}$, the proof gives the explicit estimate

$$|(1/\#Good(E, N)) \sum_{\rho \in Good(E,N)} \text{Trace}(\Lambda(\theta_{E,\rho})|$$

$$\le (1 + 1/\sqrt{\#E})(gen.rk(M) + \text{``dim''}(M))/\sqrt{\#E}.$$

CHAPTER 8

Isogenies, Connectedness, and Lie-Irreducibility

For each prime to p integer n, we have the n'th power homomorphism $[n] : G \to G$. Formation of the direct image

$$M \mapsto [n]_\star M$$

is an exact functor from $Perv$ to itself, which maps Neg to itself, $\mathcal{P}$ to itself, and which (because a homomorphism) is compatible with middle convolution:

$$[n]_\star(M \star_{mid} N) \cong ([n]_\star M) \star_{mid} ([n]_\star N).$$

So for a given object N in $\mathcal{P}_{arith}$, $[n]_\star$ allows us to view $<N>_{arith}$ as a Tannakian subcategory of $<[n]_\star N>_{arith}$, and $<N>_{geom}$ as a Tannakian subcategory of $<[n]_\star N>_{geom}$. For the fibre functor ω defined (after a choice of isomorphism $G/\overline{k} \cong \mathbb{G}_m/\overline{k}$) by

$$N \mapsto \omega(N) := H^0(\mathbb{A}^1/\overline{k}, j_{0!}N),$$

we have canonical functorial isomorphisms

$$\omega(N) = \omega([n]_\star N).$$

So with respect to these fibre functors we have inclusions of Tannakian groups

$$G_{geom,[n]_\star N} \subset G_{geom,N}$$

and

$$G_{arith,[n]_\star N} \subset G_{arith,N}.$$

Theorem 8.1. *Suppose that N in $\mathcal{P}_{geom}$ is semisimple and that n is a prime to p integer. Then $G_{geom,[n]_\star N}$ is a normal subgroup of $G_{geom,N}$. The quotient group $G_{geom,N}/G_{geom,[n]_\star N}$ is the cyclic group $\mathbb{Z}/d\mathbb{Z}$ with*

$$d := \#\{\zeta \in \mu_n(\overline{k})| \ \delta_\zeta \in <N>_{geom}\},$$

and $G_{geom,[n]_\star N}$, seen inside $G_{geom,N}$, is the fixer of the objects

$$\{\delta_\zeta \in <N>_{geom} \mid \zeta \in \mu_n(\overline{k})\}.$$

Proof. As N is semisimple, $[n]_\star N$ is also semisimple, hence $G_{geom,[n]_\star N}$ is reductive. So $G_{geom,[n]_\star N}$ is the fixer of its invariants in some representation of the ambient group $G_{geom,N}$, say corresponding to an object M in $<N>_{geom}$. Its invariants are the δ_1 subobjects of $[n]_\star M$. These are precisely the subobjects $[n]_\star\delta_\zeta$, $\zeta \in \mu_n(\overline{k})$, of $[n]_\star M$, i.e., the images by $[n]_\star$ of the subobjects δ_ζ, $\zeta \in \mu_n(\overline{k})$, of M. Thus its space of invariants in M is the largest subobject of M which is punctual and supported in $\mu_n(\overline{k})$. So the space of $G_{geom,[n]_\star N}$-invariants is a $G_{geom,N}$-stable subspace. Hence the fixer of these invariants, namely $G_{geom,[n]_\star N}$, is a normal subgroup of $G_{geom,N}$. A representation of the quotient is an object M in $<N>_{geom}$ with $[n]_\star M$ geometrically trivial, i.e., a punctual object in $<N>_{geom}$ which is supported in $\mu_n(\overline{k})$, i.e., a sum of the objects

$$\{\delta_\zeta \in <N>_{geom} \mid \zeta \in \mu_n(\overline{k})\}.$$

Thus we recover the reductive normal subgroup $G_{geom,[n]_\star N}$ of $G_{geom,N}$ as the fixer of these objects. $\square$

Recall that a representation ρ of an algebraic group G is said to be Lie-irreducible if it is both irreducible and remains irreducible when restricted to the identity component G^0 of G.

Theorem 8.2. *Suppose that N in $\mathcal{P}_{arith}$ is arithmetically semisimple and pure of weight zero (i.e., ι-pure of weight zero for every ι). Then N is geometrically Lie-irreducible (i.e., Lie-irreducible as a representation of $G_{geom,N}$) if and only if $[n]_\star N$ is geometrically irreducible for every integer $n \geq 1$ prime to p. For n_0 the order of the finite group $G_{geom,N}/G^0_{geom,N}$, we have*

$$G^0_{geom,N} = G_{geom,[n_0]_\star N}.$$

Proof. If N is geometrically Lie-irreducible, then any subgroup of finite index in $G_{geom,N}$ acts irreducibly. By the previous result, for each $n \geq 1$ prime to p the group $G_{geom,[n]_\star N}$ is of finite index, so acts irreducibly, i.e., $[n]_\star N$ is geometrically irreducible. Conversely, by Theorem 6.5, we know that $G_{geom,N}/G^0_{geom,N}$ is cyclic of some order n_0 prime to p. By Corollary 6.6, we know that $G^0_{geom,N}$ is the fixer of all punctual objects in $<N>_{geom}$. Moreover, by Theorem 6.5, the irreducible such punctual objects are precisely

$$\{\delta_\zeta \in <N>_{geom} \mid \zeta \in \overline{k}^\times\} = \{\delta_\zeta \in <N>_{geom} \mid \zeta \in \mu_{n_0}(\overline{k})\}.$$

Their fixer, by Theorem 8.1 above, is the subgroup $G_{geom,[n_0]_\star N}$. Thus we have

$$G^0_{geom,N} = G_{geom,[n_0]_\star N}.$$

This second group acts irreducibly if (and only if) $[n_0]_\star M$ is geometrically irreducible. $\square$

Corollary 8.3. *Suppose that N in $\mathcal{P}_{arith}$ is geometrically irreducible and pure of weight zero (i.e., ι-pure of weight zero for every ι). Then N is geometrically Lie-irreducible if and only if for every $a \neq 1 \in \overline{k}^\times$ the multiplicative translate $[x \mapsto ax]^\star N$ is not geometrically isomorphic to N.*

Proof. Given two semisimple objects in $\mathcal{P}_{geom}$, say $A = \bigoplus_i n_i C_i$ and $B = \bigoplus_i m_i C_i$ where the C_i are pairwise non-isomorphic, geometrically irreducible objects, and the integers n_i and m_i are ≥ 0, define the inner product

$$<A, B>_{geom} := \sum_i n_i m_i.$$

Thus a geometrically semisimple object N is geometrically irreducible if and only if $<N, N>_{geom} = 1$. Frobenius reciprocity gives, for each integer n prime to p,

$$<[n]_\star N, [n]_\star N>_{geom} = <N, [n]^\star [n]_\star N>_{geom}$$

$$= <N, \bigoplus_{\zeta \in \mu_n(\overline{k})} [x \mapsto \zeta x]^\star N>_{geom}.$$

By the previous theorem, N is geometrically Lie-irreducible if and only if $[n]_\star N$ is geometrically irreducible for every integer $n \geq 1$ prime to p, i.e., if and only if $<[n]_\star N, [n]_\star N>_{geom} = 1$ for every integer $n \geq 1$ prime to p. By Frobenius reciprocity, this holds if and only if N is not geometrically isomorphic to any nontrivial multiplicative translate of itself by a root of unity of order prime to p. But every element of $\overline{k}^\times$ is a root of unity of order prime to p. $\square$

CHAPTER 9

Autodualities and Signs

Suppose that N in $\mathcal{P}_{arith}$ is geometrically irreducible (so a fortiori arithmetically irreducible) and ι-pure of weight zero. Suppose further that N is arithmetically self-dual in $\mathcal{P}_{arith}$, i.e., that there is an arithmetic isomorphism $N \cong [x \mapsto 1/x]^*DN$, DN denoting the Verdier dual of N. This arithmetic isomorphism is then unique up to a scalar factor. It induces an autoduality on $\omega(N)$ which is respected by $G_{arith,N}$. Up to a scalar factor, this is the unique autoduality on $\omega(N)$ which is respected by $G_{arith,N}$, so it is either an orthogonal or a symplectic autoduality. We say that the duality has the sign $+1$ if it is orthogonal, and the sign -1 if it is symplectic.

Theorem 9.1. *Suppose that N in $\mathcal{P}_{arith}$ is geometrically irreducible, ι-pure of weight zero, and arithmetically self-dual. Denote by ϵ the sign of its autoduality. For variable finite extension fields E/k, we have the estimate for ϵ*

$$|\epsilon - (1/\#G(E)) \sum_{\rho \in \text{Good}(E,N)} \text{Trace}((Frob^2_{E,\rho}|\omega(N))| = O(1/\sqrt{\#E}).$$

Proof. Choose a maximal compact subgroup K of the complex reductive group $G_{arith,N}(\mathbb{C})$. For each finite extension E/k and each character ρ of $G(E)$ which is good for N, denote by $\theta_{E,\rho}$ the conjugacy class in K given by $Frob^{ss}_{E,\rho}$.

To explain the idea of the proof, first consider the special case in which we have the equality $G_{geom,N} = G_{arith,N}$. Then by Theorem 7.3, we know that as $\#E$ grows, the conjugacy classes $\theta_{E,\rho}$ become equidistributed in the space of conjugacy classes of K. As K is Zariski dense in $G_{arith,N}/\mathbb{C}$, $\omega(N)$ is K-irreducible, and K respects the autoduality. So the sign ϵ is the Frobenius-Schur indicator

$$\epsilon = \int_K \text{Trace}(k^2)dk.$$

By equidistribution, this integral is the large $\#E$ limit of the sums

$$(1/\#G(E)) \sum_{\rho \in \text{Good}(E,N)} \text{Trace}((Frob^2_{E,\rho}|\omega(N)),$$

and the proof of the equidistribution shows that the error is $O(1/\sqrt{\#E})$.

To treat the general case, we argue as follows. Because N is geometrically irreducible and arithmetically self-dual, we have, doing linear algebra in the (sense of the) Tannakian category $<N>_{arith}$,

$$End(N) \cong N \star_{mid} N \cong \mathrm{Sym}^2(N) \oplus \Lambda^2(N),$$

where End, Sym^2 and Λ^2 are all taken in the Tannakian sense. Because N is geometrically irreducible, $End(N)$ has a one-dimensional space of $G_{geom,N}$-invariants, namely the scalars, and $G_{arith,N}$ acts trivially on this space. So when we write $N \star_{mid} N$ as a sum of $G_{arith,N}$-irreducible summands, there is a unique summand which is geometrically trivial, and that summand is δ_1 itself. Every other arithmetically irreducible summand M is geometrically nontrivial. The sign ϵ is 1 if δ_1 lies in $\mathrm{Sym}^2(N)$, and it is -1 if δ_1 lies in $\Lambda^2(N)$. We have the linear algebra identity

$$\mathrm{Trace}((Frob^2_{E,\rho}|\omega(N))$$

$$= \mathrm{Trace}((Frob_{E,\rho}|\mathrm{Sym}^2(\omega(N))) - \mathrm{Trace}((Frob_{E,\rho}|\Lambda^2(\omega(N)))$$

$$= \mathrm{Trace}((Frob_{E,\rho}|\omega(\mathrm{Sym}^2(N))) - \mathrm{Trace}((Frob_{E,\rho}|\omega(\Lambda^2(N))).$$

So what we need to show is that for an arithmetically irreducible M which is ι-pure of weight zero and geometrically nontrivial, we have the estimate, as $\#E$ grows,

$$(1/\#G(E)) \sum_{good \ \rho} \mathrm{Trace}(Frob_{E,\rho}|\omega(M)) = O(1/\sqrt{\#E}).$$

This sum is within $O(1/\#E)$ of the sum

$$(1/\#G(E)) \sum_{all \ \rho} \mathrm{Trace}(Frob_E|H^0_c(G/\overline{k}, M \otimes \mathcal{L}_\rho)) = \mathrm{Trace}(Frob_{E,1}|M),$$

the last equality by orthogonality of characters.

We now distinguish two cases. If M is of the form $\mathcal{G}[1]$ for an (arithmetically irreducible, but we will not use this) middle extension sheaf $\mathcal{G}$ which is ι-pure of weight -1, then

$$\mathrm{Trace}(Frob_{E,1}|M) = -\mathrm{Trace}(Frob_{E,1}|\mathcal{G}).$$

This trace is $O(1/\sqrt{\#E})$ precisely because $\mathcal{G}$ is ι-pure of weight -1, so that its stalk at 1 (or at any other point of G(E)) is ι-mixed of weight ≤ -1.

If M is punctual, arithmetically irreducible and geometrically nontrivial, then its support is a closed point Z which is not the point 1 (lest M be geometrically trivial), and hence $\mathrm{Trace}(Frob_{E,1}|M) = 0$. $\square$

Here is a variant of the above result, with the same proof.

Theorem 9.2. *Suppose that N in $\mathcal{P}_{arith}$ is geometrically irreducible (so a fortiori arithmetically irreducible) and ι-pure of weight zero. Then we have the following results.*

(1) *If N is not geometrically self-dual, then we have the estimate, for growing finite extensions E/k,*

$$|(1/\#G(E)) \sum_{good\ \rho} \mathrm{Trace}(Frob^2_{E,\rho}|\omega(N))| = O(1/\sqrt{\#E}).$$

(2) *If N is geometrically self-dual, with sign of autoduality ϵ, then $N \star_{mid} N$ contains exactly one arithmetically irreducible summand which is geometrically trivial, of the form $\alpha^{deg} \otimes \delta_1$ for some unitary scalar α. Every other arithmetically irreducible summand is geometrically nontrivial, and we have the estimate, for growing finite extensions E/k,*

$$|\epsilon \alpha^{deg(E/k)} - (1/\#G(E)) \sum_{good\ \rho} \mathrm{Trace}(Frob^2_{E,\rho}|\omega(N))| = O(1/\sqrt{\#E}).$$

(3) *In the situation of (2), if $\alpha = 1$, then N is arithmetically self-dual.*

Remark 9.3. In the situation (2) above, we can approximately recover the unitary scalar α by taking the ratio of the sums

$$(1/\#G(E)) \sum_{good\ \rho} \mathrm{Trace}(Frob^2_{E,\rho}|\omega(N))$$

for two finite extensions E/k of large degrees $n+1$ and n. Once we know α approximately, we can then determine ϵ approximately, and hence exactly, given that it is ± 1.

CHAPTER 10

A First Construction of Autodual Objects

These constructions are based on evaluating the sum

$$(1/\#G(E)) \sum_{\rho \in \mathrm{Good}(E,N)} \mathrm{Trace}(Frob_{E,\rho}^2 | \omega(N))$$

$$= (1/\#G(E)) \sum_{\rho \in \mathrm{Good}(E,N)} \mathrm{Trace}(Frob_{E_2} | H_c^0(G/\overline{k}, N \otimes \mathcal{L}_\rho))$$

more or less precisely. As always, this sum is within $O(1/\#E)$ of the sum

$$(1/\#G(E)) \sum_{all\ \rho \in G(E)^\vee} \mathrm{Trace}(Frob_{E_2} | H_c^0(G/\overline{k}, N \otimes \mathcal{L}_\rho)),$$

which is in turn equal, by the Lefschetz Trace formula [**Gr-Rat**], to

$$(1/\#G(E)) \sum_{all\ \rho \in G(E)^\vee} \sum_{t \in G(E_2)} \rho(\mathrm{Norm}_{E_2/E}(t)) \mathrm{Trace}(Frob_{E_2,t} | N).$$

This sum, by orthogonality, is

$$\sum_{t \in G(E_2) | \mathrm{Norm}_{E_2/E}(t) = 1} \mathrm{Trace}(Frob_{E_2,t} | N).$$

We begin with a geometrically irreducible middle extension sheaf $\mathcal{F}$ on G/k which is ι-pure of weight zero, and which is not geometrically an $\mathcal{L}_\chi$. Thus $\mathcal{F}(1/2)[1]$ is a geometrically irreducible object in $\mathcal{P}_{arith}$. Its dual in $\mathcal{P}_{arith}$ is $[x \mapsto 1/x]^\star \overline{\mathcal{F}}(1/2)[1]$, for $\overline{\mathcal{F}}$ the linear dual middle extension sheaf. Via ι, $\mathcal{F}$ and $\overline{\mathcal{F}}$ have complex conjugate trace functions; this holds by ι-purity on the dense open set where $\mathcal{F}$ is lisse, and then on all of G by a result of Gabber [**Fuj-Indep**, Thm. 3], cf. also [**Ka-MMP**, proof of 1.8.1 (i)].

Theorem 10.1. *Suppose that the tensor product sheaf*

$$\mathcal{G} := \mathcal{F} \otimes [x \mapsto 1/x]^\star \overline{\mathcal{F}}$$

is itself a middle extension sheaf; this is automatic if either $\mathcal{F}$ is lisse on G or if the finite set S of points of $G(\overline{k})$ at which $\mathcal{F}$ is ramified is disjoint from the set $1/S$ of its inverses. Suppose further that $\mathcal{G}$ is *geometrically irreducible, and not geometrically isomorphic to an $\mathcal{L}_\chi$. Then $\mathcal{G}(1/2)[1]$ is a geometrically irreducible object of $\mathcal{P}_{arith}$ which is symplectically self-dual.*

Proof. It is obvious from the description of the dual of an object N of $\mathcal{P}_{arith}$ as $[x \mapsto 1/x]^*DN$ that $\mathcal{G}(1/2)[1]$ is arithmetically self-dual. The sign ϵ of the autoduality is given approximately by the sum

$$\sum_{t \in G(E_2)|\mathrm{Norm}_{E_2/E}(t)=1} \mathrm{Trace}(Frob_{E_2,t}|N).$$

$$= - \sum_{t \in G(E_2)|\mathrm{Norm}_{E_2/E}(t)=1} \mathrm{Trace}(Frob_{E_2,t}|\mathcal{G}(1/2)).$$

Because we are taking Frobenii over E_2, the $1/2$ Tate twist pulls out a factor $1/\#E$, so this last sum is

$$= (-1/\#E) \sum_{t \in G(E_2)|\mathrm{Norm}_{E_2/E}(t)=1} \mathrm{Trace}(Frob_{E_2,t}|\mathcal{G})$$

$$= (-1/\#E) \sum_{t \in G(E_2)|\mathrm{Norm}_{E_2/E}(t)=1} \mathrm{Trace}(Frob_{E_2,t}|\mathcal{F})\mathrm{Trace}(Frob_{E_2,1/t}|\overline{\mathcal{F}}).$$

The key observation is that for $t \in G(E_2)$ with $\mathrm{Norm}_{E_2/E}(t) = 1$, we have $1/t = \sigma(t)$ for σ the nontrivial element in $Gal(E_2/E)$. Thus

$$\mathrm{Trace}(Frob_{E_2,1/t}|\overline{\mathcal{F}}) = \mathrm{Trace}(Frob_{E_2,\sigma(t)}|\overline{\mathcal{F}})$$

for such a t. On the other hand, since $\overline{\mathcal{F}}$ starts life on G/k, we have

$$\mathrm{Trace}(Frob_{E_2,\sigma(t)}|\overline{\mathcal{F}}) = \mathrm{Trace}(Frob_{E_2,t}|\overline{\mathcal{F}}),$$

which is in turn equal to

$$\overline{\mathrm{Trace}(Frob_{E_2,t}|\mathcal{F})}.$$

So our sum is

$$= (-1/\#E) \sum_{t \in G(E_2)|\mathrm{Norm}_{E_2/E}(t)=1} |\mathrm{Trace}(Frob_{E_2,t}|\mathcal{F})|^2,$$

which is negative or zero. But for large $\#E$ this sum is approximately the sign ϵ, which is ± 1, so for large $\#E$ the sum cannot vanish, so must be strictly negative. Hence the sign ϵ, which is ± 1, must be -1. $\square$

CHAPTER 11

A Second Construction of Autodual Objects

In this construction, we work on the split form $\mathbb{G}_m/k$, $Spec(k[x, 1/x])$. We begin with a geometrically irreducible lisse sheaf $\mathcal{F}$ on an open dense set $U \subset \mathbb{G}_m$ which is ι-pure of weight zero and which is self-dual: $\mathcal{F} \cong \overline{\mathcal{F}}$.

Denote by d the rank of $\mathcal{F}$. We view $\mathcal{F}|U$ as a d-dimensional representation ρ of $\pi_1^{arith}(U)$, toward either the orthogonal group $O(d)/\overline{\mathbb{Q}_\ell}$, if the autoduality is orthogonal, or toward the symplectic group $Sp(d)/\overline{\mathbb{Q}_\ell}$ if the autoduality is symplectic (which forces d to be even). We denote by $G_{geom, \mathcal{F}}$ the Zariski closure of the image $\rho(\pi_1^{geom}(U))$ of the **geometric** fundamental group.[1]

We have a finite morphism

$$\pi : \mathbb{G}_m[1/(x^2 + 1)] \to \mathbb{G}_m, \quad x \mapsto x + 1/x.$$

Then $\pi^\star \mathcal{F}$ is lisse and ι-pure of weight zero on some open set $j : V \subset \mathbb{G}_m$.

Theorem 11.1. *For $\mathcal{F}$ as above, consider the middle extension sheaf $\mathcal{G} := j_\star \pi^\star \mathcal{F}$ on $\mathbb{G}_m/k$. Suppose in addition the following three conditions hold.*

(1) *If $d = 1$, $\mathcal{G}$ is not geometrically a Kummer sheaf $\mathcal{L}_\chi$.*
(2) *If the autoduality of $\mathcal{F}$ is orthogonal, then $d \neq 2$ and $G_{geom, \mathcal{F}}$ is either $O(d)$ or $SO(d)$.*
(3) *If the autoduality of $\mathcal{F}$ is symplectic, $G_{geom, \mathcal{F}}$ is $Sp(d)$.*

Then $N := \mathcal{G}(1/2)[1]$ lies in $\mathcal{P}_{arith}$ and is ι-pure of weight zero. It is geometrically irreducible and arithmetically self-dual. The sign of its autoduality is opposite to that of $\mathcal{F}$.

Proof. The lisse sheaf $\mathcal{F}$ on U is geometrically Lie-irreducible, because $G^0_{geom, \mathcal{F}}$, which is either $SO(d)$, $d \neq 2$, or $Sp(d)$, acts irreducibly in its standard representation. Therefore $\pi^\star \mathcal{F}$, or indeed any pullback of $\mathcal{F}$ by a finite morphism, remains geometrically Lie-irreducible. Thus N is perverse, ι-pure of weight zero., and geometrically irreducible, so

---

[1]See the notational caution at the very end of the Introduction.

by (1) is a geometrically irreducible object of $\mathcal{P}_{arith}$. It is arithmetically self-dual, because isomorphic to both its Verdier dual (thanks to the autoduality of $\mathcal{F}$) and to its pullback by multiplicative inversion (thanks to having been pulled back by π). It remains to determine the sign ϵ_N of its autoduality.

The idea is quite simple. Let us denote by $\epsilon_{\mathcal{F}}$ the sign of the autoduality of $\mathcal{F}$. Then $\epsilon_{\mathcal{F}}$ is the large $\#E$ limit of the sums

$$(1/\#E) \sum_{s \in U(E)} \mathrm{Trace}(Frob^2_{E,s}|\mathcal{F}) = (1/\#E) \sum_{s \in U(E)} \mathrm{Trace}(Frob_{E_2,s}|\mathcal{F}),$$

cf. [**Ka-GKM**, 4.2]. On the other hand, ϵ_N is the large $\#E$ limit of the sums

$$= - \sum_{t \in G(E_2)|\mathrm{Norm}_{E_2/E}(t)=1} \mathrm{Trace}(Frob_{E_2,t}|\mathcal{G}(1/2))$$

$$= (-1/\#E) \sum_{t \in G(E_2)|\mathrm{Norm}_{E_2/E}(t)=1} \mathrm{Trace}(Frob_{E_2,t}|\mathcal{G}).$$

Now for any $t \in G(E_2)$ with $t^2 + 1 \neq 0$ and such that $t + 1/t$ lies in U, we have

$$\mathrm{Trace}(Frob_{E_2,t}|\mathcal{G}) = \mathrm{Trace}(Frob_{E_2,t+1/t}|\mathcal{F}).$$

So ϵ_N is the large $\#E$ limit of the sums

$$= (-1/\#E) \sum_{t \in G(E_2)|\mathrm{Norm}_{E_2/E}(t)=1, t^2+1\neq0, t+1/t\in U} \mathrm{Trace}(Frob_{E_2,t+1/t}|\mathcal{F}).$$

For $t \in G(E_2)$ with $\mathrm{Norm}_{E_2/E}(t) = 1$, $t + 1/t$ is just $\mathrm{Trace}(t)$. With the exception of the points ± 1, every point $t \in G(E_2)$ with $\mathrm{Norm}_{E_2/E}(t) = 1$ has degree two over E, so is a root of a quadratic polynomial $T^2 - sT + 1 \in E[T]$. Here $s = t + 1/t$. Conversely, an irreducible quadratic polynomial of the form $T^2 - sT + 1 \in E[T]$ has two roots, t and $1/t$ in $G(E_2)$ with $\mathrm{Norm}_{E_2/E}(t) = \mathrm{Norm}_{E_2/E}(1/t) = 1, t+1/t = s$. In other words, the set of $t \in G(E_2)$ with $\mathrm{Norm}_{E_2/E}(t) = 1$, $t \neq \pm 1$, is a double covering, by $t \mapsto t + 1/t = \mathrm{Trace}(t)$, of the set of $s \in E$ such that the quadratic polynomial $T^2 - sT + 1$ is E-irreducible. Thus this last sum is within $O(1\#E)$ of the sum

$$= 2(-1/\#E) \sum_{s \in U(E)|\ T^2-sT+1\ irred./E} \mathrm{Trace}(Frob_{E_2,s}|\mathcal{F}).$$

We will show that the sums whose large $\#E$ limit is $\epsilon_{\mathcal{F}}$, namely

$$(1/\#E) \sum_{s \in U(E)} \mathrm{Trace}(Frob_{E_2,s}|\mathcal{F}),$$

are within $O(1/\sqrt{\#E})$ of the sums

$$= 2(1/\#E) \sum_{s \in U(E)| \ T^2 - sT + 1 \ irred./E} \text{Trace}(Frob_{E_2,s}|\mathcal{F})$$

whose large $\#E$ limit is $-\epsilon_N$.

We first treat the case when k has odd characteristic. Shrinking U if necessary, we may suppose that neither 2 nor -2 lies in U. Now $T^2 - sT + 1$ is E-irreducible if and only if its discriminant $s^2 - 4$ is a nonzero nonsquare in E. Denoting by $\chi_{2,E}$ the quadratic character of $E^\times$, we then have, for $s \in U(E)$,

$$1 - \chi_{2,E}(s^2 - 4) = 0$$

if $s^2 - 4$ is a square, and

$$1 - \chi_{2,E}(s^2 - 4) = 2$$

if $s^2 - 4$ is a nonsquare.

So the sums whose large $\#E$ limit is $-\epsilon_N$ are

$$(1/\#E) \sum_{s \in U(E)} \text{Trace}(Frob_{E_2,s}|\mathcal{F})(1 - \chi_{2,E}(s^2 - 4)).$$

Hence we are reduced to showing that the sums

$$(1/\#E) \sum_{s \in U(E)} \text{Trace}(Frob_{E_2,s}|\mathcal{F})\chi_{2,E}(s^2 - 4)$$

are $O(1/\sqrt{\#E})$. To see this, we make use of the linear algebra identity

$$\text{Trace}(Frob_{E_2,s}|\mathcal{F}) = \text{Trace}(Frob_{E,s}|\text{Sym}^2(\mathcal{F})) - \text{Trace}(Frob_{E,s}|\Lambda^2(\mathcal{F})).$$

So it suffices to prove that both of the sums

$$(1/\#E) \sum_{s \in U(E)} \text{Trace}(Frob_{E,s}|\text{Sym}^2(\mathcal{F}))\chi_{2,E}(s^2 - 4)$$

$$= (1/\#E) \sum_{s \in U(E)} \text{Trace}(Frob_{E,s}|\text{Sym}^2(\mathcal{F}) \otimes \mathcal{L}_{\chi_2(s^2-4)}))$$

and

$$(1/\#E) \sum_{s \in U(E)} \text{Trace}(Frob_{E,s}|\Lambda^2(\mathcal{F}))\chi_{2,E}(s^2 - 4)$$

$$= (1/\#E) \sum_{s \in U(E)} \text{Trace}(Frob_{E,s}|\Lambda^2(\mathcal{F}) \otimes \mathcal{L}_{\chi_2(s^2-4)}))$$

are $O(1/\sqrt{\#E})$.

If the arithmetic autoduality of $\mathcal{F}$ has $G_{geom,\mathcal{F}} = Sp(d)$, then $\text{Sym}^2(\mathcal{F})$ is (both arithmetically and) geometrically irreducible, and it has rank > 1. Therefore its tensor product with any lisse rank one sheaf, here $\mathcal{L}_{\chi_2(s^2-4)}$, remains (both arithmetically and) geometrically irreducible. Therefore its $H_c^2(U/\overline{k}, \mathrm{Sym}^2(\mathcal{F}) \otimes \mathcal{L}_{\chi_2(s^2-4)}))$ vanishes. Because $\mathcal{F}$ is ι-pure of weight zero, so is $\mathrm{Sym}^2(\mathcal{F}) \otimes \mathcal{L}_{\chi_2(s^2-4)}$, and hence by Deligne [**De-Weil II**, 3.3.1] its H_c^1 is ι-mixed of weight ≤ 1. By the Lefschetz Trace formula [**Gr-Rat**] and the vanishing of the H_c^2,

$$(1/\#E) \sum_{s \in U(E)} \mathrm{Trace}(Frob_{E,s}|\mathrm{Sym}^2(\mathcal{F}))\chi_{2,E}(s^2 - 4)$$

$$= -(1/\#E)\mathrm{Trace}(Frob_E|H_c^1(U/\overline{k}, \mathrm{Sym}^2(\mathcal{F}) \otimes \mathcal{L}_{\chi_2(s^2-4)}))$$

is $O(1/\sqrt{\#E})$. And the sheaf $\Lambda^2(\mathcal{F})$ is either the constant sheaf $\overline{\mathbb{Q}_\ell}$ if $d = 2$ or, if $d \geq 4$, the direct sum

$$\Lambda^2(\mathcal{F}) = \overline{\mathbb{Q}_\ell} \oplus \mathcal{H},$$

with $\mathcal{H}$ an arithmetically and geometrically irreducible lisse sheaf of rank > 1. Again $\mathcal{H} \otimes \mathcal{L}_{\chi_2(s^2-4)}$ is ι-pure of weight zero and has vanishing H_c^2, and an H_c^1 which is ι-mixed of weight ≤ 1. So the sum

$$(1/\#E) \sum_{s \in U(E)} \mathrm{Trace}(Frob_{E,s}|\mathcal{H} \otimes \mathcal{L}_{\chi_2(s^2-4)}))$$

is $O(1/\sqrt{\#E})$. The final term is

$$(1/\#E) \sum_{s \in U(E)} \mathrm{Trace}(Frob_{E,s}|\mathcal{L}_{\chi_2(s^2-4)}),$$

which again is $O(1/\sqrt{\#E})$ because $\mathcal{L}_{\chi_2(s^2-4)}$ is geometrically nontrivial.

If the arithmetic autoduality of $\mathcal{F}$ has $G_{geom,\mathcal{F}}$ containing $SO(d)$ with $d \neq 2$, we argue as follows. We first treat separately the case $d = 1$. Then $\mathcal{F}$ is a lisse sheaf of rank one whose trace function takes values in ± 1. Therefore $\Lambda^2(\mathcal{F}) = 0$, and $\mathrm{Sym}^2(\mathcal{F}) = \mathcal{F}^{\otimes 2}$ is the constant sheaf. So the sum we must estimate is just

$$(1/\#E) \sum_{s \in U(E)} \mathrm{Trace}(Frob_{E,s}|\mathcal{L}_{\chi_2(s^2-4)}),$$

which as noted above is $O(1/\sqrt{\#E})$.

If $d \geq 3$, the argument is essentially identical to the argument in the symplectic case, except that now it is $\Lambda^2(\mathcal{F})$ which is arithmetically and geometrically irreducible of rank > 1, and it is $\mathrm{Sym}^2(\mathcal{F})$ which admits a direct sum decomposition

$$\mathrm{Sym}^2(\mathcal{F}) = \overline{\mathbb{Q}_\ell} \oplus \mathcal{H},$$

with $\mathcal{H}$ an arithmetically and geometrically irreducible lisse sheaf of rank > 1.

It remains to treat the case of characteristic 2, where we no longer have the discriminant to tell us when $T^2 - sT + 1$ is E-irreducible. In characteristic 2 we consider directly the finite étale covering of $\mathbb{G}_m$ by $\mathbb{G}_m \setminus \{1\}$ given by $\pi : t \mapsto t + 1/t := s$. This extends to a finite étale covering of $\mathbb{P}^1 \setminus \{0\}$ by $\mathbb{P}^1 \setminus \{1\}$. Making the change of variable $t = u/(u+1)$, we readily compute

$$1/s = 1/(t + 1/t) = 1/(u/(u+1) + (u+1)/u) = u^2 - u.$$

Thus in characteristic two, the role of $\mathcal{L}_{\chi_2(s^2-4)}$ is now played by the Artin-Schreier sheaf $\mathcal{L}_{\psi(1/s)}$. With this change, we just repeat the proof from odd characteristic.   □

Here is a slight generalization of the previous result, where we relax the hypotheses on the group $G_{geom,\mathcal{F}}$ attached to the geometrically irreducible lisse sheaf $\mathcal{F}$ on an open dense set $U \subset \mathbb{G}_m$ which is ι-pure of weight zero and which is self-dual: $\mathcal{F} \cong \overline{\mathcal{F}}$. We make the following two hypotheses.

(1) The identity component $G^0_{geom,\mathcal{F}}$ acts irreducibly in its given d-dimensional representation ρ.

(2) In the representation $\rho \otimes \rho$ corresponding to $\mathcal{F} \otimes \mathcal{F}$, the space of invariants under $G_{geom,\mathcal{F}}$ is one-dimensional, and every other irreducible constituent of $\rho \otimes \rho$ for the action of $G_{geom,\mathcal{F}}$ has dimension > 1.

These conditions are automatically satisfied in the symplectic case when $G_{geom,\mathcal{F}} = Sp(d)$, and in the orthogonal case when $d \neq 2$ and $G_{geom,\mathcal{F}}$ contains $SO(d)$. But they are also satisfied if $G_{geom,\mathcal{F}}$ receives an $SL(2)$ such that $\rho|SL(2)$ is irreducible. For then $\rho|SL(2)$ must be $\mathrm{Sym}^{d-1}(std_2)$ as $SL(2)$-representation, and one knows that

$$\mathrm{Sym}^{d-1}(std_2) \otimes \mathrm{Sym}^{d-1}(std_2) = \bigoplus_{r=0}^{d-1} \mathrm{Sym}^{2d-2-2r}(std_2)$$

as $SL(2)$-representation. For later use, let us recall that in the world of $GL(2)$-representations, we have

$$\mathrm{Sym}^{d-1}(std_2) \otimes \mathrm{Sym}^{d-1}(std_2) = \bigoplus_{r=0}^{d-1} \mathrm{Sym}^{2d-2-2r}(std_2) \otimes det^{\otimes r}.$$

In fact, one has the more precise decompositions

$$\mathrm{Sym}^2(\mathrm{Sym}^{d-1}(std_2)) = \bigoplus_{r=0}^{[(d-1)/2]} \mathrm{Sym}^{2d-2-4r}(std_2) \otimes det^{\otimes 2r}$$

and

$$\Lambda^2(\mathrm{Sym}^{d-1}(std_2)) = \bigoplus_{r=0}^{[(d-2)/2]} \mathrm{Sym}^{2d-4-4r}(std_2) \otimes det^{\otimes 2r+1}.$$

Theorem 11.2. *For $\mathcal{F}$ as above, satisfying the hypotheses (1) and (2), consider the middle extension sheaf $\mathcal{G} := j_{V\star}\pi^\star\mathcal{F}$ on $\mathbb{G}_m/k$. Then $N := \mathcal{G}(1/2)[1]$ lies in $\mathcal{P}_{arith}$ and is ι-pure of weight zero. It is geometrically irreducible and arithmetically self-dual. The sign of its autoduality is opposite to that of $\mathcal{F}$.*

Proof. Repeat the proof of Theorem 11.1. In the symplectic case, $\Lambda^2(\mathcal{F})$ has an invariant corresponding to the symplectic form, so we have a direct sum decomposition $\Lambda^2(\mathcal{F}) = \mathcal{H} \oplus \overline{\mathbb{Q}_\ell}$. By hypothesis (2), each $G_{geom,\mathcal{F}}$-irreducible constituent of $\mathcal{H}$ has dimension ≥ 2, and each $G_{geom,\mathcal{F}}$-irreducible constituent of $\mathrm{Sym}^2(\mathcal{F})$ has dimension ≥ 2. These conditions ensure the vanishings of the various H_c^2 in the proof. In the orthogonal case, reverse the roles of Sym^2 and Λ^2. $\square$

CHAPTER 12

The Previous Construction in the Nonsplit Case

In this construction, we work on the nonsplit form G/k. Denoting by k_2/k the unique quadratic extension inside $\overline{k}$, recall that for any k-algebra A, $G(A)$ is the group of elements $t \in (A \otimes_k k_2)^\times$ with $\text{Norm}_{A \otimes_k k_2/A}(t) = 1$. Thus $G(A) \subset A \otimes_k k_2$. We have the trace map $\text{Trace}_{A \otimes_k k_2/A} : A \otimes_k k_2 \to A$. Restricting it to $G(A)$, we get, for any k-algebra A, a map $\text{Trace} : G(A) \to A$, in other words a k-morphism $\text{Trace} : G/k \to \mathbb{A}^1/k$, i.e., a function Trace on G/k.

A basic observation is that the function Trace is invariant under inversion: for $t \in G(A)$, $\text{Trace}(t) = \text{Trace}(1/t)$. Indeed for $t \in G(A)$, we claim that $\text{Trace}(t) = t + 1/t$ when we view $A \subset A \otimes_k k_2$. To see this,[1] present k_2 as $k[u]/(u^2 + au + b)$ with $u^2 + au + b \in k[u]$ quadratic (and irreducible, but we will not use this, or any other property, of the quadratic polynomial in question). Then an element $t \in A \otimes_k k_2/A$, written as $t = x + yu$ with $x, y \in A$, acts by multiplication on $A \otimes_k k_2$ by the two by two matrix, with respect to the basis $1, u$, $(x, y, -by, x - ay)$. So we have

$$\text{Norm}_{A \otimes_k k_2/A}(x + yu) = x^2 - axy + by^2 = (x + yu)((x - ay) - yu),$$

$$\text{Trace}_{A \otimes_k k_2/A}(x + yu) = 2x - ay.$$

So if $t = x + yu \in G(A)$, then $1/t = (x - ay) - yu$, and so we find $\text{Trace}(t) = t + 1/t \in A \otimes_k k_2$, as asserted. Moreover, if $u^2 + au + b \in k[u]$ is **reducible** and has distinct roots (so that we are dealing with the split form), say $u^2 + au + b = (u - \alpha)(u - \beta)$ with $\alpha, \beta \in k$, $\alpha \neq \beta$, then $t = x + yu \mapsto (X, Y) := (x + y\alpha, x + y\beta)$ is an isomorphism of G/k with $\mathbb{G}_m/k$ as the locus $XY = 1$, where the trace is the function $(X, 1/X) \mapsto X + 1/X$.

Observe that the trace morphism $\text{Trace} : G/k \to \mathbb{A}^1/k$ is a finite morphism. To see this, we may extend scalars from k to k_2, where G becomes $\mathbb{G}_m$ and Trace becomes the map $t \mapsto t + 1/t$.

---

[1] Another way to see this is to use the fact that k_2/k is galois of degree 2. For σ the nontrivial element of the Galois group, we have the automorphism $\sigma_A := id \otimes \sigma$ of $A \otimes_k k_2$, and for $t \in A \otimes_k k_2$, we have $\text{Norm}_{A \otimes_k k_2/A}(t) = t\sigma_A(t)$, $\text{Trace}_{A \otimes_k k_2/A}(t) = t + \sigma_A(t)$. Thus if $t \in G(A)$, then $\sigma_A(t) = 1/t$, and hence $\text{Trace}_{A \otimes_k k_2/A}(t) = t + 1/t$.

We now give the construction. We begin with a geometrically irreducible lisse sheaf $\mathcal{F}$ on an open dense set $U \subset \mathbb{A}^1$ which is ι-pure of weight zero and which is self-dual: $\mathcal{F} \cong \overline{\mathcal{F}}$. Denote by d the rank of $\mathcal{F}$. We view $\mathcal{F}|U$ as a representation ρ of $\pi_1^{arith}(U)$, toward either the orthogonal group $O(d)/\overline{\mathbb{Q}_\ell}$, if the autoduality is orthogonal, or toward the symplectic group $Sp(d)/\overline{\mathbb{Q}_\ell}$ if the autoduality is symplectic (which forces d to be even). We denote by $G_{geom,\mathcal{F}}$ the Zariski closure of the image $\rho(\pi_1^{geom}(U))$ of the **geometric** fundamental group.

As noted above, the trace gives a finite morphism

$$\text{Trace} : G \to \mathbb{A}^1.$$

Then $\text{Trace}^\star \mathcal{F}$ is lisse and ι-pure of weight zero on some open set $j : V \subset G$.

Theorem 12.1. *For $\mathcal{F}$ as above, consider the middle extension sheaf $\mathcal{G} := j_\star \text{Trace}^\star \mathcal{F}$ on G/k. Suppose in addition the following two conditions hold.*

(1) *The identity component $G^0_{geom,\mathcal{F}}$ acts irreducibly in its given d-dimensional representation ρ.*

(2) *In the representation $\rho \otimes \rho$ corresponding to $\mathcal{F} \otimes \mathcal{F}$, the space of invariants under $G_{geom,\mathcal{F}}$ is one-dimensional, and every other irreducible constituent of $\rho \otimes \rho$ for the action of $G_{geom,\mathcal{F}}$ has dimension > 1.*

Then $N := \mathcal{G}(1/2)[1]$ lies in $\mathcal{P}_{arith}$ and is ι-pure of weight zero. It is geometrically irreducible and arithmetically self-dual. The sign of its autoduality is opposite to that of $\mathcal{F}$.

Proof. That N lies in $\mathcal{P}_{arith}$, is ι-pure of weight zero, geometrically irreducible and arithmetically self-dual is proven exactly as in the proof of Theorem 11.1, with the π there replaced by Trace. To compute the sign ϵ_N, we can take the large $\#E$ limit over finite extension fields E/k of **even** degree. This reduces us to the split case, already treated in Theorem 11.1. □

CHAPTER 13

Results of Goursat-Kolchin-Ribet Type

Suppose we are given some number $r \geq 2$ of objects $N_1, N_2, ..., N_r$ in $\mathcal{P}_{arith}$ of some common "dimension" $d \geq 1$. Suppose they are all ι-pure of weight zero, geometrically irreducible, and arithmetically self-dual, all with the same sign of duality.

Theorem 13.1. *Suppose that $d \geq 2$ is even, that each N_i is symplectically self-dual, and that for each $i = 1, ..., r$, we have $G_{geom,N_i} = G_{arith,N_i} = Sp(d)$. Suppose further that for $i \neq j$, there is no geometric isomorphism between N_i and N_j and there is no geometric isomorphism between N_i and $[x \mapsto -x]^\star N_j$. Then the direct sum $\oplus_i N_i$ has*

$$G_{geom,\oplus_i N_i} = G_{arith,\oplus_i N_i} = \prod_{i=1}^r Sp(d).$$

Proof. We apply the Goursat-Kolchin-Ribet theorem [**Ka-ESDE**, 1.8.2] to the group $G_{geom,\oplus_i N_i}$ and its representations V_i corresponding to the N_i. In order to show that $G_{geom,\oplus_i N_i} = \prod_i Sp(d)$, it suffices to show that for $i \neq j$, there is no geometric isomorphism between N_i and $N_j \star_{mid} L$, for any one-dimensional object L of $<\oplus_i N_i>_{geom}$. In fact, we will show that, under the hypotheses of the theorem, there is none for any one-dimensional object of $\mathcal{P}_{geom}$. Suppose there were. Both N_j and $N_j \star_{mid} L \cong N_i$ are symplectic representations, with image $Sp(d)$. But the only scalars in $Sp(d)$ are ± 1. Therefore $L^{\otimes 2}$ in the Tannakian sense, i.e., $L \star_{mid} L$, is geometrically trivial. One knows [**Ka-ESDE**, 8.5.3] that the only one-dimensional objects in $\mathcal{P}_{geom}$ are delta objects δ_a for some $a \in \overline{k}^\times$ and multiplicative translates of hypergeometric sheaves placed in degree -1. But such hypergeometric objects are of infinite order, as middle self-convolution simply produces other such objects, of larger and larger generic rank. Therefore our L must be some δ_a. Because $L \star_{mid} L$ is geometrically trivial, we conclude that $a = \pm 1$. But middle convolution with δ_a is just multiplicative translation by a. Therefore we have $G_{geom,\oplus_i N_i} = \prod_i Sp(d)$. Since in any case we have $G_{geom,\oplus_i N_i} \subset G_{arith,\oplus_i N_i} \subset \prod_i Sp(d)$, we get the asserted equality

$$G_{geom, \oplus_i N_i} = G_{arith, \oplus_i N_i} = \prod_i Sp(d).$$

$\square$

Theorem 13.2. *Suppose that $d \geq 6$ is even, $d \neq 8$, that each N_i is orthogonally self-dual, and that for each $i = 1, ..., r$, we have $G_{geom, N_i} = SO(d)$. Suppose further that for $i \neq j$, there is no geometric isomorphism between N_i and N_j and there is no geometric isomorphism between N_i and $[x \mapsto -x]^* N_j$. Then the direct sum $\oplus_i N_i$ has*

$$G_{geom, \oplus_i N_i} = \prod_{i=1}^r SO(d) \subset G_{arith, \oplus_i N_i} \subset \prod_{i=1}^r O(d).$$

If in addition we have $G_{geom, N_i} = G_{arith, N_i} = SO(d)$ for every i, then

$$G_{geom, \oplus_i N_i} = G_{arith, \oplus_i N_i} = \prod_{i=1}^r SO(d).$$

Proof. The proof, via the Goursat-Kolchin-Ribet theorem [**Ka-ESDE,** 1.8.2], is identical to the previous one. $\square$

Theorem 13.3. *Suppose that $d \geq 3$ is odd, that each N_i is orthogonally self-dual, and that for each $i = 1, ..., r$, we have $G_{geom, N_i} = SO(d)$. Suppose further that for $i \neq j$, there is no geometric isomorphism between N_i and N_j. Then the direct sum $\oplus_i N_i$ has*

$$G_{geom, \oplus_i N_i} = \prod_{i=1}^r SO(d) \subset G_{arith, \oplus_i N_i} \subset \prod_{i=1}^r O(d).$$

If in addition we have $G_{geom, N_i} = G_{arith, N_i} = SO(d)$ for every i, then

$$G_{geom, \oplus_i N_i} = G_{arith, \oplus_i N_i} = \prod_{i=1}^r SO(d).$$

Proof. The proof, again via the Goursat-Kolchin-Ribet theorem [**Ka-ESDE** 1.8.2] is even simpler in this case, because for d odd, $SO(d)$ contains no scalars other than 1, so L can only be δ_1. $\square$

In the orthogonal case, the Goursat-Kolchin-Ribet theorem [**Ka-ESDE,** 1.8.2] gives the following less precise version of these last two theorems.

Theorem 13.4. *Suppose that either $d \geq 6$ is even, $d \neq 8$, or $d \geq 3$ is odd. Suppose that each N_i is orthogonally self-dual, and that for each $i = 1, ..., r$, we have $SO(d) \subset G_{geom, N_i}$. Suppose further that for $i \neq j$, there is no geometric isomorphism between N_i and N_j and there is no* *geometric isomorphism between* N_i *and* $[x \mapsto -x]^\star N_j$. *Then the direct sum* $\oplus_i N_i$ *has*

$$\prod_{i=1}^{r} SO(d) \subset G_{geom, \oplus_i N_i} \subset G_{arith, \oplus_i N_i} \subset \prod_{i=1}^{r} O(d).$$

We end this chapter with the case of $SL(d)$.

Theorem 13.5. *Suppose that for each* $i = 1, ..., r$, *we have* $SL(d) \subset G_{geom, N_i}$. *Denote by* $N_i^\vee$ *the dual in the Tannakian sense of* N_i. *Suppose further that for every one-dimensional object* $L \in \mathcal{P}_{geom}$, *and for* $i \neq j$, *there is no geometric isomorphism between* N_i *and* $N_j \star_{mid} L$ *nor between* N_i *and* $N_j^\vee \star_{mid} L$. *Then the direct sum* $\oplus_i N_i$ *has*

$$\prod_{i=1}^{r} SL(d) \subset G_{geom, \oplus_i N_i} \subset G_{arith, \oplus_i N_i} \subset \prod_{i=1}^{r} GL(d).$$

CHAPTER 14

The Case of $SL(2)$; the Examples of Evans and Rudnick

In treating both of these examples, as well as all the examples to come, we will use the Euler-Poincaré formula, cf. [**Ray**, Thm. 1] or [**Ka-GKM**, 2.3.1] or [**Ka-SE**, 4.6, (v) atop p. 113] or [**De-ST**, 3.2.1], to compute the "dimension" of the object N in question.

Let us briefly recall the general statement of the Euler-Poincaré formula, and then specialize to the case at hand. Let X be a projective, smooth, nonsingular curve over an algebraically closed field $\overline{k}$ in which ℓ is invertible, $U \subset X$ a dense open set in X, and $V \subset U$ a dense open set in U. Let $\mathcal{G}$ be a constructible $\overline{\mathbb{Q}}_\ell$-sheaf on U which is lisse on V of rank $r := gen.rk.(\mathcal{G})$. We view $\mathcal{G}|V$ as a representation of $\pi_1(V)$. For each point $x \in (X \setminus V)(\overline{k})$, we restrict this representation to the inertia group $I(x)$ at x. Its Swan conductor gives a nonnegative integer $Swan_x(\mathcal{G})$, cf. [**Ka-GKM**, 1.5-1.10], which vanishes if and only if $\mathcal{G}|V$ is tamely ramified at x. For each point $u \in (U \setminus V)(\overline{k})$, we have the integer $drop_u(\mathcal{G}) := gen.rk.(\mathcal{G}) - dim(\mathcal{G}_u)$. We denote by $\chi(U)$ the Euler characteristic of U. Thus if X has genus g, then $\chi(U) = 2 - 2g - \#(X \setminus U)(\overline{k})$. The Euler-Poincaré formula states that $\chi_c(U, \mathcal{G}) = \chi(U, \mathcal{G})$ is equal to

$$\chi(U)gen.rk.(\mathcal{G}) - \sum_{x \in (X \setminus V)(\overline{k})} Swan_x(\mathcal{G}) - \sum_{u \in (U \setminus V)(\overline{k})} drop_u(\mathcal{G}).$$

When U is $\mathbb{G}_m$, whose χ vanishes, and we place $\mathcal{G}$ in degree -1, this becomes

$$\chi_c(\mathbb{G}_m, \mathcal{G}[1]) = Swan_0(\mathcal{G}) + Swan_\infty(\mathcal{G}) + \sum_{u \in (\mathbb{G}_m \setminus V)(\overline{k})} [drop_u(\mathcal{G}) + Swan_u(\mathcal{G})].$$

The term inside the square brackets, $drop_u(\mathcal{G}) + Swan_u(\mathcal{G})$, is called the "total drop" of $\mathcal{G}$ at u. Using this terminology, the formula becomes

$$\chi_c(\mathbb{G}_m, \mathcal{G}[1]) = Swan_0(\mathcal{G}) + Swan_\infty(\mathcal{G}) + \sum_{u \in (\mathbb{G}_m \setminus V)(\overline{k})} TotalDrop_u(\mathcal{G}).$$

We now turn to the key result of this chapter.

Theorem 14.1. *Suppose N in $\mathcal{P}_{arith}$ is pure of weight zero, geometrically irreducible, and arithmetically self-dual of "dimension" two. Then the following conditions are equivalent.*

(1) *N is geometrically Lie-irreducible.*
(2) *N is not geometrically isomorphic to any nontrivial multiplicative translate $[x \mapsto ax]^*N$, $a \neq 1$, of itself.*
(3) *N is symplectically self-dual, and we have $G_{geom,N} = G_{arith,N} = SL(2)$.*

Proof. The equivalence of (1) and (2) was proven in Corollary 8.3. If (3) holds, then $G^0_{geom} = SL(2)$ acts irreducibly in its standard representation, hence (1) holds. Conversely, suppose (1) holds. We will show that (3) holds. The autoduality on N is either orthogonal or symplectic. We first show it cannot be orthogonal. Indeed, if it were orthogonal, we would have $G_{geom,N} \subset G_{arith,N} \subset O(2)$, with $G_{geom,N}$ a Lie-irreducible subgroup of $O(2)$. But there are no such subgroups, because $SO(2)$ has index two in $O(2)$ and is abelian. Therefore the autoduality must be symplectic. So we have inclusions $G_{geom,N} \subset G_{arith,N} \subset SL(2)$, and hence it suffices to show that $G_{geom,N} = SL(2)$. But the only irreducible (in the standard representation) subgroups of $SL(2)/\overline{\mathbb{Q}}_\ell$ are $SL(2)$ itself, the normalizer $N(T)$ of a torus T, and some finite subgroups. Of these, $SL(2)$ is the only one that acts Lie-irreducibly.   □

With this result in hand, it is a simple matter to treat the examples of Evans and of Rudnick. We begin by treating the example of Evans.

Theorem 14.2. *For ψ a nontrivial additive character of k, and N the object $\mathcal{L}_{\psi(x-1/x)}(1/2)[1]$ in $\mathcal{P}_{arith}$ on $\mathbb{G}_m/k := Spec(k[x, 1/x])$, we have $G_{geom,N} = G_{arith,N} = SL(2)$.*

Proof. The lisse sheaf $\mathcal{L}_{\psi(x-1/x)}$ is pure of weight zero. It is wildly ramified at both 0 and ∞, with Swan conductor 1 at each, so is not geometrically isomorphic to an $\mathcal{L}_\chi$. Thus N is a geometrically irreducible object of $\mathcal{P}_{arith}$, pure of weight zero. Its dimension "$\dim$"$N := \chi(\mathbb{G}_m/\overline{k}, N)$ is given by the Euler-Poincaré formula,

$$\chi(\mathbb{G}_m/\overline{k}, N) = Swan_0(\mathcal{L}_{\psi(x-1/x)}) + Swan_\infty(\mathcal{L}_{\psi(x-1/x)}) = 1 + 1 = 2.$$

Writing $\mathcal{L}_{\psi(x-1/x)} = \mathcal{L}_{\psi(x)} \otimes \mathcal{L}_{\psi(-1/x)}$, we see by Theorem 10.1 that N is symplectically self-dual. The multiplicative translate of N by $a \in \overline{k}^\times$ is $\mathcal{L}_{\psi(ax-1/ax)}(1/2)[1]$. This is geometrically isomorphic to N if and only if $\mathcal{L}_{\psi(ax-1/ax)}$ is geometrically isomorphic to $\mathcal{L}_{\psi(x-1/x)}$, i.e., if and only if their ratio $\mathcal{L}_{\psi(ax-1/ax)} \otimes (\mathcal{L}_{\psi(x-1/x)})^{-1} = \mathcal{L}_{\psi((a-1)x+(1-1/a)/x)}$ is geometrically trivial. But for $a \neq 1$, this ratio is itself wildly ramified at both 0 and ∞. Therefore N is geometrically Lie-irreducible, and we conclude by the previous result. $\qquad\square$

Here is a strengthening of this result, using Theorem 12.1.

Theorem 14.3. *Let $c_1, ..., c_r$ be $r \geq 2$ elements of $k^\times$ whose squares are distinct: for $i \neq j$, $c_i \neq \pm c_j$. Denote by N_i the object $\mathcal{L}_{\psi(c_i(x-1/x))}(1/2)[1]$ on $\mathbb{G}_m/k := Spec(k[x, 1/x])$. Then we have $G_{geom, \oplus_i N_i} = G_{arith, \oplus_i N_i} = \prod_{i=1}^r SL(2)$.*

Proof. We must show that for $i \neq j$, $\mathcal{L}_{\psi(c_i(x-1/x))}$ is not geometrically isomorphic to either $\mathcal{L}_{\psi(c_j(x-1/x))}$ or to $[x \mapsto -x]^\star \mathcal{L}_{\psi(c_j(x-1/x))} = \mathcal{L}_{\psi(-c_j(x-1/x))}$. As in the proof above, the ratio is $\mathcal{L}_{\psi((-c_i \pm c_j)(x-1/x))}$. This is wildly ramified at both 0 and ∞, because for $i \neq j$, $-c_i \pm c_j \neq 0$. $\qquad\square$

Here is a further strengthening of this result, again using Theorem 12.1. Its very formulation is based on the fact that for a given object $N \in \mathcal{P}_{arith}$ and a given Kummer sheaf $\mathcal{L}_\chi$, the functor $M \mapsto M \otimes \mathcal{L}_\chi$ induces a Tannakian isomorphism of $<N>_{arith}$ with $<N \otimes \mathcal{L}_\chi>_{arith}$, and of $<N>_{geom}$ with $<N \otimes \mathcal{L}_\chi>_{geom}$. In particular N and $N \otimes \mathcal{L}_\chi$ have the "same" groups G_{geom} and the "same" groups G_{arith} as each other.

Theorem 14.4. *Let $c_1, ..., c_r$ be $r \geq 1$ elements of $k^\times$ whose squares are distinct: for $i \neq j$, $c_i \neq \pm c_j$. Let $\chi_1, ..., \chi_s$ be $s \geq 1$ distinct characters of $k^\times$. Denote by $N_{i,j}$ the object $\mathcal{L}_{\psi(c_i(x-1/x))} \otimes \mathcal{L}_{\chi_j}(1/2)[1]$ on $\mathbb{G}_m/k := Spec(k[x, 1/x])$. Then we have $G_{geom, \oplus_{i,j} N_{i,j}} = G_{arith, \oplus_{i,j} N_{i,j}} = \prod_{i=1}^r \prod_{j=1}^s SL(2)$.*

Proof. We must show that for $(i, j) \neq (a, b)$, $\mathcal{L}_{\psi(c_i(x-1/x))} \otimes \mathcal{L}_{\chi_j}$ is not geometrically isomorphic to either $\mathcal{L}_{\psi(c_a(x-1/x))} \otimes \mathcal{L}_{\chi_b}$ or to $[x \mapsto -x]^\star \mathcal{L}_{\psi(c_a(x-1/x))} \otimes \mathcal{L}_{\chi_b} \overset{geom}{\cong} \mathcal{L}_{\psi(-c_a(x-1/x))} \otimes \mathcal{L}_{\chi_b}$. [Recall that, geometrically, Kummer sheaves $\mathcal{L}_\chi$ are invariant under multiplicative translation.] If $i \neq a$, both ratios are wildly ramified at both 0 and ∞, just as in the proof of the previous result. If $i = a$ but $j \neq b$, then the ratio is either wildly ramified at both 0 and ∞, or it is $\mathcal{L}_{\chi_j/\chi_b}$, which is not geometrically constant. $\qquad\square$

We now turn to the example of Rudnick in the split case.

Theorem 14.5. *Suppose that k has odd characteristic. We work on $\mathbb{G}_m/k := Spec(k[x, 1/x])$. For ψ a nontrivial additive character of k, form the lisse sheaf $\mathcal{L}_{\psi((x+1)/(x-1))}$ on $U := \mathbb{G}_m \setminus \{1\}$. For $j : U \subset \mathbb{G}_m$ the inclusion, we have $j_! \mathcal{L}_{\psi((x+1)/(x-1))} = j_\star \mathcal{L}_{\psi((x+1)/(x-1))}$. Form the* object $N := j_\star \mathcal{L}_{\psi((x+1)/(x-1))}(1/2)[1]$ on $\mathbb{G}_m/k$. *Then N is a geometrically irreducible two-dimensional object of $\mathcal{P}_{arith}$ which is pure of weight zero and arithmetically self-dual. It has $G_{geom,N} = G_{arith,N} = SL(2)$.*

Proof. The sheaf $\mathcal{L}_{\psi((x+1)/(x-1))}$ is wildly ramified at 1, so it is not geometrically isomorphic to an $\mathcal{L}_\chi$. Thus N is a geometrically irreducible object of $\mathcal{P}_{arith}$, which is pure of weight zero. It is lisse at both 0 and ∞. Its only singularity in $\mathbb{G}_m$ is at 1, where its Swan conductor is 1. So the Euler-Poincaré formula, cf. [**Ray**] or [**Ka-GKM**, 2.3.1], shows that its dimension is two. It is symplectically self-dual, by Theorem 10.1 applied to the lisse sheaf $\mathcal{F} := \mathcal{L}_{\psi((1/2)(x+1)/(x-1))}$ on U. And N is not geometrically isomorphic to a multiplicative translate of itself by any $a \neq 1$, because 1 is the unique point of $\mathbb{G}_m(\overline{k})$ at which N is not lisse. $\qquad\square$

Here is a strengthening of this result, using Theorem 12.1.

Theorem 14.6. *Let $c_1, ..., c_r$ be $r \geq 2$ distinct elements of $k^\times$. Form the object $N_i := j_\star \mathcal{L}_{\psi(c_i(x+1)/(x-1))}(1/2)[1]$ on $\mathbb{G}_m/k$. Then we have $G_{geom,\oplus_i N_i} = G_{arith,\oplus_i N_i} = \prod_{i=1}^r SL(2)$.*

Proof. We must show that for $i \neq j$, $\mathcal{L}_{\psi(c_i(x+1)/(x-1))}$ is not geometrically isomorphic to either $\mathcal{L}_{\psi(c_j(x+1)/(x-1))}$ or to $[x \mapsto -x]^\star \mathcal{L}_{\psi(c_j(x+1)/(x-1))}$. The second isomorphism is impossible because the source is singular only at $1 \in \mathbb{G}_m$ while the target is singular only at $-1 \in \mathbb{G}_m$. For $i \neq j$, $\mathcal{L}_{\psi(c_i(x+1)/(x-1))}$ is not geometrically isomorphic to $\mathcal{L}_{\psi(c_j(x+1)/(x-1))}$ because their ratio, $\mathcal{L}_{\psi((c_j-c_i)(x+1)/(x-1))}$, is wildly ramified at 1. $\qquad\square$

Here is a further strengthening, again using Theorem 12.1.

Theorem 14.7. *Let $c_1, ..., c_r$ be $r \geq 1$ distinct elements of $k^\times$. Let $\chi_1, ..., \chi_s$ be $s \geq 1$ distinct characters of $k^\times$. Denote by $N_{a,b}$ the object $N_{a,b} := j_\star \mathcal{L}_{\psi(c_a(x+1)/(x-1))} \otimes \mathcal{L}_{\chi_b}(1/2)[1]$ on $\mathbb{G}_m/k$. Then we have $G_{geom,\oplus_{a,b} N_{a,b}} = G_{arith,\oplus_{a,b} N_{a,b}} = \prod_{a=1}^r \prod_{b=1}^s SL(2)$.*

Proof. All the objects $N_{a,b}$ have the point 1 as their unique singularity, so just as in the argument above it suffices to show that for $(a,b) \neq (c,d)$, $N_{a,b}$ is not geometrically isomorphic to $N_{c,d}$. If $a \neq b$, then just as above the ratio is wildly ramified at 1. If $a = b$ but $c \neq d$, the ratio is $\mathcal{L}_{\chi_d/\chi_c}$, which is not geometrically constant. $\qquad\square$

We conclude this chapter with the example of Rudnick in the nonsplit case. Here k has odd characteristic. Completing the square, we can present k_2 as $k[u]/(u^2+b)$ with $b \in k^\times$ and $u^2 + b \in k[u]$ quadratic (and irreducible, but we will not use this). For any k-algebra A, we write elements of $A \otimes_k k_2$ as $x + yu$, with $x, y \in A$.

Then $G/k = Spec(k[x,y]/(x^2 + by^2 - 1))$. The group law is
$$(x,y)(s,t) := (xs - byt, xt + ys),$$
the identity is $(1,0)$, and inversion is
$$[inv]^*(x,y) = (x,-y).$$
On the open set $G[1/(x-1)]$ where $x-1$ is invertible, we have the function $-by/(x-1)$, which changes sign under inversion:
$$[inv]^*(-by/(x-1)) = by/(x-1).$$
In the split case, i.e., when $-b = c^2$ for some $c \in k^\times$, then $(x,y) \mapsto t := x + cy$ is an isomorphism $G \cong \mathbb{G}_m$. Then we readily calculate[1]
$$-by/(x-1) = c(t+1)/(t-1).$$

Theorem 14.8. *Suppose that k has odd characteristic. For ψ a nontrivial additive character of k, form the lisse sheaf $\mathcal{L}_{\psi(-by/(x-1))}$ on $U := G[1/(x-1)]$. For $j : U \subset G$ the inclusion, we have $j_! \mathcal{L}_{\psi(-by/(x-1))} = j_* \mathcal{L}_{\psi(-by/(x-1))}$. Form the object $N := j_* \mathcal{L}_{\psi(-by/(x-1))}(1/2)[1]$ on G/k. Then N is a geometrically irreducible two-dimensional object of $\mathcal{P}_{arith}$ which is pure of weight zero and arithmetically self-dual. It has $G_{geom,N} = G_{arith,N} = SL(2)$.*

Proof. That $j_* \mathcal{L}_{\psi(-by/(x-1))}$ is not geometrically a Kummer sheaf $\mathcal{L}_\chi$ is a geometric statement, already proven in the split case, as is the fact that N is geometrically irreducible two-dimensional object of $\mathcal{P}_{arith}$. That it is pure of weight zero is obvious from its definition. That it is arithmetically symplectically self-dual results from Theorem 10.1, applied to $\mathcal{L}_{\psi(-by/2(x-1))}$. The rest of the proof is the same as in the split case. $\square$

For the sake of completeness, here are the strengthenings in the nonsplit case, with the same proofs as in the split case.

Theorem 14.9. *Let $c_1, ..., c_r$ be $r \geq 2$ distinct elements of $k^\times$. Form the object $N_i := j_* \mathcal{L}_{\psi(-bc_iy/(x-1))}(1/2)[1]$ on G/k. Then we have $G_{geom,\oplus_i N_i} = G_{arith,\oplus_i N_i} = \prod_{i=1}^r SL(2)$.*

Theorem 14.10. *Let $c_1, ..., c_r$ be $r \geq 1$ distinct elements of $k^\times$. Let $\chi_1, ..., \chi_s$ be $s \geq 1$ distinct characters of $G(k)$. Denote by $N_{i,j}$ the object $N_{i,j} := j_* \mathcal{L}_{\psi(-bc_iy/(x-1))} \otimes \mathcal{L}_{\chi_j}(1/2)[1]$ on G/k. Then we have $G_{geom,\oplus_{i,j} N_{i,j}} = G_{arith,\oplus_{i,j} N_{i,j}} = \prod_{i=1}^r \prod_{j=1}^s SL(2)$.*

---

[1]Indeed, $t = x + cy$, $1/t = x - cy$, so $2cy = t - 1/t = (t+1)(t-1)/t$, $2x - 2 = t + 1/t - 2 = (t-1)^2/t$, hence $cy/(x-1) = (t+1)/(t-1)$.

CHAPTER 15

Further $SL(2)$ Examples, Based on the Legendre Family

In this chapter, we suppose that k has odd characteristic. We begin with the Legendre family of elliptic curves over the λ line, given in $\mathbb{P}^2 \times \mathbb{A}^1$ by the equation

$$Y^2 Z = X(X - Z)(X - \lambda Z).$$

For π its projection onto $\mathbb{A}^1$, we define

$$Leg := R^1 \pi_! \overline{\mathbb{Q}_\ell}.$$

Thus Leg is lisse of rank two and pure of weight one (Hasse's theorem [**Ha-Ell**, page 205]) outside of 0 and 1. For $j : \mathbb{A}^1 \setminus \{0, 1\} \subset \mathbb{A}^1$ the inclusion we have $Leg = j_* j^* Leg$. One knows [**De-Weil II**, 3.5.5] that the geometric monodromy group of the lisse sheaf $j^* Leg$ is $SL(2)$. Its local monodromy at both 0 and 1 is a single unipotent block $Unip(2)$; its local monodromy at ∞ is $\mathcal{L}_{\chi_2(\lambda)} \otimes Unip(2)$, cf. [**Ka-Sar**, 10.1.7].

We will also make use of the quadratic twist

$$TwLeg := j_* j^* (\mathcal{L}_{\chi_2(1-\lambda)} \otimes Leg).$$

Over $\mathbb{A}^1 \setminus \{0, 1\}$, this quadratic twist $TwLeg$ is the $R^1 \pi_! \overline{\mathbb{Q}_\ell}$ for the family

$$Y^2 Z = (1 - \lambda)X(X - Z)(X - \lambda Z).$$

When -1 is a square in k, we have an arithmetic isomorphism over $\mathbb{A}^1 \setminus \{0, 1\}$,

$$TwLeg \cong [\lambda \mapsto 1/\lambda]^* TwLeg,$$

indeed the two families are isomorphic. The geometric monodromy group of the lisse sheaf $j^* TwLeg$ is $SL(2)$. The local monodromy of $TwLeg$ at both 0 and ∞ is a single unipotent block $Unip(2)$, while its local monodromy at 1 is $\mathcal{L}_{\chi_2(1-\lambda)} \otimes Unip(2)$.

Theorem 15.1. *Suppose k has odd characteristic, and -1 is a square in k. Consider the object $N := TwLeg(1)[1] \in \mathcal{P}_{arith}$. This object is a geometrically Lie-irreducible two-dimensional object of $\mathcal{P}_{arith}$ which is pure of weight zero and arithmetically self-dual. It has $G_{geom,N} = G_{arith,N} = SL(2)$.*

Proof. That N is geometrically irreducible results from the fact that j^*TwLeg has geometric monodromy group $SL(2)$, which is connected and acts irreducibly in its standard representation. That N is pure of weight zero is an instance of Hasse's theorem [**Ha-Ell**, page 205]. That N is geometrically Lie-irreducible, i.e., not geometrically isomorphic to any nontrivial multiplicative translate, results from the fact that 1 is its only singularity in $\mathbb{G}_m$. That it is two-dimensional is immediate from the Euler-Poincaré formula (as it is tame at both 0 and ∞, and has a drop of two at 1), cf. [**Ray**] or [**Ka-GKM**, 2.3.1]. That it is arithmetically self-dual results from fact that $TwLeg$ has an integer (and hence real) valued trace function, and the isomorphism $TwLeg \cong [\lambda \mapsto 1/\lambda]^*TwLeg$. The result now follows from Theorem 14.1.   □

We also have the following strengthening, as always using Theorem 12.1.

Theorem 15.2. *Suppose k has odd characteristic, and -1 is a square in k. Let $\chi_1, ..., \chi_s$ be $s \geq 2$ distinct characters of $k^\times$. Consider the objects $N_i := TwLeg \otimes \mathcal{L}_{\chi_i}(1)[1] \in \mathcal{P}_{arith}$. Then $G_{geom, \oplus_i N_i} = G_{arith, \oplus_i N_i} = \prod_{i=1}^{s} SL(2)$.*

Proof. We must show that for $i \neq j$, $TwLeg \otimes \mathcal{L}_{\chi_i}$ is not geometrically isomorphic to either $TwLeg \otimes \mathcal{L}_{\chi_j}$ or to $[\lambda \mapsto -\lambda]^*(TwLeg \otimes \mathcal{L}_{\chi_j})$. The latter is impossible, because $TwLeg \otimes \mathcal{L}_{\chi_i}$ has its unique singularity in $\mathbb{G}_m$ at 1. The former is impossible, because already the local monodromies at 0, namely $\mathcal{L}_{\chi_i} \otimes Unip(2)$ and $\mathcal{L}_{\chi_j} \otimes Unip(2)$, are not geometrically isomorphic.   □

We can also work with the object $\mathrm{Sym}^2(Leg)$, which, we remark in passing, is the same as $\mathrm{Sym}^2(TwLeg)$, because the quadratic twist disappears after forming Sym^2. The lisse sheaf $j^*\mathrm{Sym}^2(Leg)$ is pure of weight two, and its geometric monodromy group is $SO(3)$. Its local monodromies at 0, 1, and ∞ are each $Unip(3)$, a single unipotent Jordan block of dimension three. We have an arithmetic isomorphism

$$\mathrm{Sym}^2(Leg) \cong [\lambda \mapsto 1/\lambda]^*\mathrm{Sym}^2(Leg).$$

Theorem 15.3. *Suppose k has odd characteristic. Consider the object $N := \mathrm{Sym}^2(Leg)(3/2)[1] \in \mathcal{P}_{arith}$. This object is a geometrically Lie-irreducible two-dimensional object of $\mathcal{P}_{arith}$ which is pure of weight zero and arithmetically self-dual. It has $G_{geom,N} = G_{arith,N} = SL(2)$.*

Proof. That N is geometrically irreducible results from the fact that $\mathrm{Sym}^2(Leg) = j_*j^*\mathrm{Sym}^2(Leg)$ and that $j^*\mathrm{Sym}^2(Leg)$ has its geometric monodromy group $SO(3)$, which is irreducible in its standard representation. It is pure of weight zero, by Hasse [**Ha-Ell**, page 205]. The fact that $\mathrm{Sym}^2(Leg)$ has 1 as its unique singularity shows that it is isomorphic to no nontrivial multiplicative translate of itself, so is a geometrically Lie-irreducible object of $\mathcal{P}_{arith}$. That N is self-dual results from the fact that $\mathrm{Sym}^2(Leg)$ has an integer (and hence real) valued trace function, and the isomorphism $TwLeg \cong [\lambda \mapsto 1/\lambda]^{-1}TwLeg$. That it is two-dimensional is immediate from the Euler-Poincaré formula [**Ray**] (as it is tame at both 0 and ∞, and has a drop of two at 1).The result now follows from Theorem 14.1. $\qquad\square$

Exactly as in the case of $TwLeg$, we have the following strengthening.

Theorem 15.4. *Suppose k has odd characteristic. Let $\chi_1, ..., \chi_s$ be $s \geq 2$ distinct characters of $k^\times$. Consider the objects $N_i := \mathrm{Sym}^2(Leg) \otimes \mathcal{L}_{\chi_i}(3/2)[1] \in \mathcal{P}_{arith}$. Then $G_{geom,\oplus_i N_i} = G_{arith,\oplus_i N_i} = \prod_{i=1}^s SL(2)$.*

Proof. The proof is nearly identical to that of Theorem 15.2; one has only to replace $Unip(2)$ by $Unip(3)$. $\qquad\square$

CHAPTER 16

Frobenius Tori and Weights; Getting Elements of G_{arith}

In this chapter, we work on $\mathbb{G}_m/k$. We consider an arithmetically semisimple object $N \in \mathcal{P}_{arith}$ which is pure of weight zero. We assume it is of the form $\mathcal{G}[1]$, with $\mathcal{G}$ a middle extension sheaf. Thus for some open set $j : U \subset \mathbb{G}_m$, we have $\mathcal{G} = j_\star\mathcal{F}$, for $\mathcal{F} := j^\star\mathcal{G}$ a lisse sheaf on U which is pure of weight -1 and arithmetically semisimple, and having no geometric constituent isomorphic to (the restriction to $U_{\overline{k}}$ of) a Kummer sheaf. Recall that Deligne's fibre functor is (for $j_0 : \mathbb{G}_m \subset \mathbb{A}^1$ the inclusion)

$$M \mapsto \omega(N) := H^0(\mathbb{A}^1 \otimes_k \overline{k}, j_{0!}N).$$

The action of $Frob_k$ on the restriction to $<N>_{arith}$ of this fibre functor gives us an element $Frob_{k,1} \in G_{arith,N}$. Now view $G_{arith,N}$ as a subgroup of $GL(\omega(N))$. Then $Frob_{k,1}$ is the action of $Frob_k$ on the cohomology group

$$\omega(N) := H^0(\mathbb{A}^1 \otimes_k \overline{k}, j_{0!}N) = H^1(\mathbb{A}^1 \otimes_k \overline{k}, j_{0!}\mathcal{G}).$$

We now recall the relation of the absolute values of the eigenvalues of $Frob_{k,1}$ to the local monodromies of $\mathcal{F}$ at the two points 0 and ∞. To do this, we introduce some ad hoc notation. We denote by $\mathcal{F}(0)$ (respectively by $\mathcal{F}(\infty)$) the $I(0)$ (respectively the $I(\infty)$) -representation attached to $\mathcal{F}$. We separate it as the direct sum of its tame and wild parts. We then isolate, in the tame part, the summand which is unipotent, which we denote $\mathcal{F}(0)^{unip}$ (respectively by $\mathcal{F}(\infty)^{unip}$). We denote by $d_0 \geq 0$ (respectively by $d_\infty \geq 0$) the number of Jordan blocks in $\mathcal{F}(0)^{unip}$ (respectively in $\mathcal{F}(\infty)^{unip}$). We write each as the sum of its Jordan blocks.

$$\mathcal{F}(0)^{unip} = \oplus_{i=1}^{d_0} Unip(e_i),$$
$$\mathcal{F}(\infty)^{unip} = \oplus_{j=1}^{d_\infty} Unip(f_j).$$

The following theorem is a spelling out of some of the results of Deligne's Weil II [**De-Weil II**].

Theorem 16.1. *The action of $Frob_{k,1}$ on $\omega(N) := H^0(\mathbb{A}^1 \otimes_k \overline{k}, j_{0!}N) = H^1(\mathbb{A}^1 \otimes_k \overline{k}, j_{0!}\mathcal{G})$ has exactly d_0 eigenvalues of weight < 0; their weights* *are* $-e_1, ..., -e_{d_0}$. *It has exactly* d_∞ *eigenvalues of weight* > 0; *their weights are* $f_1, ..., f_{d_\infty}$. *All other eigenvalues, if any, have weight zero.*

Proof. By the main theorem of Deligne's Weil II in the case of curves [**De-Weil II**, 3.2.3], we know that $H^1(\mathbb{P}^1 \otimes_k \overline{k}, j_{0*}j_{\infty*}\mathcal{G})$ is pure of weight zero. Our cohomology group $\omega(N) = H^1(\mathbb{A}^1 \otimes_k \overline{k}, j_{0!}\mathcal{G})$ is the group $H^1(\mathbb{P}^1 \otimes_k \overline{k}, j_{0!}Rj_{\infty*}\mathcal{G})$. We have a short exact sequence of sheaves on $\mathbb{P}^1$, with punctual third term,

$$0 \to j_{0!}j_{\infty*}\mathcal{G} \to j_{0*}j_{\infty*}\mathcal{G} \to \mathcal{F}(0)^{I(0)} \to 0$$

and a short exact sequence of perverse sheaves on $\mathbb{P}^1$, with punctual third term,

$$0 \to j_{0!}j_{\infty*}\mathcal{G}[1] \to j_{0!}Rj_{\infty*}\mathcal{G}[1] \to H^1(I(\infty), \mathcal{F}(\infty)) \to 0.$$

Because $\mathcal{G} := j_*\mathcal{F}$ has no geometrically constant constituents, the long exact cohomology sequences give short exact sequences

$$0 \to \mathcal{F}(0)^{I(0)} \to H^1(\mathbb{P}^1 \otimes_k \overline{k}, j_{0!}j_{\infty*}\mathcal{G}) \to H^1(\mathbb{P}^1 \otimes_k \overline{k}, j_{0*}j_{\infty*}\mathcal{G}) \to 0,$$

and

$$0 \to H^1(\mathbb{P}^1 \otimes_k \overline{k}, j_{0!}j_{\infty*}\mathcal{G}) \to \omega(N) \to H^1(I(\infty), \mathcal{F}(\infty)) \to 0.$$

Thus $\omega(N)$ is a successive extension of the three groups

$$H^1(I(\infty), \mathcal{F}(\infty)), \quad H^1(\mathbb{P}^1 \otimes_k \overline{k}, j_{0*}j_{\infty*}\mathcal{G}), \quad \mathcal{F}(0)^{I(0)}.$$

That $\mathcal{F}(0)^{I(0)}$ has weights $-e_1, ..., -e_{d_0}$ is [**De-Weil II**, 1.6.14.2-3 and 1.8.4], cf. [**Ka-GKM**, 7.0.7]. That $H^1(I(\infty), \mathcal{F}(\infty))$ has weights $f_1, ..., f_{d_\infty}$ is the dual (remember that $\mathcal{F}$ is pure of weight -1) statement. Because $H^1(\mathbb{P}^1 \otimes_k \overline{k}, j_{0*}j_{\infty*}\mathcal{G})$ is pure of weight zero, the impurities in $\omega(N)$ are as asserted. $\qquad\square$

The following corollary will soon play a crucial role.

Corollary 16.2. *We have the following results.*

(1) *Suppose that $\mathcal{F}(0)^{unip}$ is a single Jordan block $Unip(e)$, some $e \geq 1$, and that $\mathcal{F}(\infty)^{unip} = 0$. Then $G_{arith,N}$ contains, in a suitable basis of $\omega(N)$, the torus consisting of all diagonal elements of the form $Diag(x, 1, 1, ..., 1)$.*

(2) *Suppose that $\mathcal{F}(0)^{unip}$ is a single Jordan block $Unip(e)$, some $e \geq 1$, and that $\mathcal{F}(\infty)^{unip}$ is a single Jordan block $Unip(e)$ of the same size. Then $G_{arith,N}$ contains, in a suitable basis of $\omega(N)$, the torus consisting of all diagonal elements of the form $Diag(x, 1/x, 1, 1, ..., 1)$.*

Proof. Since $Frob_{k,1}$ lies in $G_{arith,N}$, so does its semisimplification in the sense of Jordan decomposition. This semisimplification, in a suitable basis, is the element $Diag(\alpha_1, ..., \alpha_n)$, $n = \dim(\omega(N))$, consisting of the eigenvalues α_i of $Frob_{k,1}$. The Zariski closure of the group generated by this element, Serre's Frobenius torus [**Se-Let**], is defined by all monomial equations

$$\prod_i X_i^{a_i} = 1,$$

as $(a_1, ..., a_n)$ runs over all vectors in $\mathbb{Z}^n$ such that

$$\prod_i \alpha_i^{a_i} = 1,$$

cf. [**Chev-TGL II**, Chap.2, Section &13, Prop.3]. Let us call such vectors "equation vectors." Taking absolute values via any ι, we see that any equation vector $(a_1, ..., a_n) \in \mathbb{Z}^n$ satisfies

$$\prod_i |\alpha_i|^{a_i} = 1.$$

In case (1), all but one of the α_i has $|\alpha_i| = 1$, and the remaining one, say α_1, has $|\alpha_i| = 1/\sqrt{\#k^e}$. So we conclude that $a_1 = 0$ in any equation vector. Hence any element $Diag(x, 1, 1, ..., 1)$ satisfies all the equations.

In case (2), after renumbering we have $|\alpha_1| = 1/\sqrt{\#k^e}$, $|\alpha_2| = \sqrt{\#k^e}$, and $|\alpha_i| = 1$ for $i \geq 3$. So here we infer that any equation vector has $a_1 = a_2$, and hence that any element $Diag(x, 1/x, 1, 1, ..., 1)$ satisfies all the equations. $\square$

We end this chapter with another application of Theorem 16.1. We have already used, in the proof of Corollary 6.3, the fact that the only one-dimensional objects in $\mathcal{P}_{geom}$ are the delta objects δ_a for $a \in \overline{k}^\times$ and multiplicative translates of shifted hypergeometric sheaves

$$\mathcal{H}(\psi; \chi_1, ..., \chi_n; \rho_1, ..., \rho_m)[1]$$

of type (n, m) where $\text{Max}(n, m) \geq 1$ and no χ_i is a ρ_j. We have already characterized the delta sheaves as those of finite order, or, in the case when our one-dimensional object lies in $\mathcal{P}_{arith}$, as those with no bad characters. Using Theorem 16.1, we can give a recipe for the one-dimensional objects in $\mathcal{P}_{arith}$ of infinite order, in terms of weight losses and weight gains.

Corollary 16.3. *Let L in $\mathcal{P}_{arith}$ be a one-dimensional object on $\mathbb{G}_m/k$ which is pure of weight zero. If L has no bad characters, then L is δ_a*

for some $a \in k^{\times}$. If L has bad characters, then L is geometrically iso-morphic to a multiplicative translate of a shifted hypergeometric sheaf $\mathcal{H}(\psi; \chi_1, ..., \chi_n; \rho_1, ..., \rho_m)((n+m)/2)[1]$ with $\mathrm{Max}(n,m) \geq 1$ and such that no χ_i is a ρ_j. A character χ occurs among the χ_i if and only if, over some finite extension E/k where χ is defined, $Frob_{E,\overline{\chi}}|\omega(L)$ has weight $-w_{\chi} < 0$, and in that case χ occurs exactly w_{χ} times among the χ_i. A character ρ occurs among the ρ_j if and only if, over some finite extension E/k where ρ is defined, $Frob_{E,\overline{\rho}}|\omega(L)$ has weight $w_{\rho} > 0$, and in that case χ occurs exactly w_{ρ} times among the ρ_j.

Proof. The case of no bad characters is simply "mise pour mémoire." The case where there are bad characters is immediate from Theorem 16.1, given the known local monodromies at 0 and ∞ of hypergeometric sheaves, cf. [**Ka-ESDE**, 8.4.2, (7)-(9)]. $\qquad\qquad\square$

CHAPTER 17

$GL(n)$ **Examples**

Here we work on either the split or the nonsplit form. We begin with a lisse sheaf $\mathcal{F}$ on a dense open set $j : U \subset G$ which is geometrically irreducible, pure of weight zero, and not geometrically isomorphic to (the restriction to U of) any Kummer sheaf $\mathcal{L}_\chi$. We denote by $\mathcal{G} := j_\star \mathcal{F}$ its middle extension to G. Then the object $N := \mathcal{G}(1/2)[1] \in \mathcal{P}_{arith}$ is pure of weight zero and geometrically irreducible.

Theorem 17.1. *Suppose that N is not geometrically isomorphic to any nontrivial multiplicative translate of itself. Suppose further that for one of the two possible geometric isomorphisms $G/\overline{k} \cong \mathbb{G}_m/\overline{k}$, $\mathcal{F}(0)^{unip}$ is a single Jordan block $Unip(e)$ for some $e \geq 1$, and $\mathcal{F}(\infty)^{unip} = 0$. For $n := \dim(\omega(N))$ we have*

$$G_{geom,N} = G_{arith,N} = GL(n).$$

Proof. We have a priori inclusions

$$G_{geom,N} \subset G_{arith,N} \subset GL(n),$$

so it suffices to prove that $G_{geom,N} = GL(n)$. So we may extend scalars if necessary from k to its quadratic extension k_2, and reduce to the case where G is $\mathbb{G}_m$, $\mathcal{F}(0)^{unip}$ is a single Jordan block $Unip(e)$, and $\mathcal{F}(\infty)^{unip} = 0$. Then by the previous chapter, $G_{arith,N}$ contains, in a suitable basis of $\omega(N)$, the torus $Diag(x, 1, 1, ...1)$.

The hypothesis that N is not geometrically isomorphic to any nontrivial multiplicative translate of itself insures that N is geometrically Lie-irreducible, i.e., that G^0_{geom} is an irreducible subgroup of $GL(n)$. Hence G^0_{geom} is the almost product of its center, which by irreducibility consists entirely of scalars, with its derived group (:=commutator subgroup) $G^{0,der}_{geom}$, which is a connected semisimple group which also must act irreducibly.

Because $G_{geom,N}$ is a normal subgroup of $G_{arith,N}$, so also its intrinsic subgroup $G^{0,der}_{geom,N}$ is a normal subgroup of $G_{arith,N}$. Therefore the Lie algebra $Lie(G^{0,der}_{geom,N})$ is a semisimple irreducible Lie subalgebra of $End(\omega(N))$ which is normalized by $G_{arith,N}$. In particular, $Lie(G^{0,der}_{geom,N})$ is normalized by all elements $Diag(x, 1, 1, ...1)$. Such elements are pseudoreflections, with determinant equal to x. Take $x \neq \pm 1$. Our semisimple irreducible Lie subalgebra of $End(\omega(N))$ is normalized by a pseudoreflection of determinant not ± 1. The only such Lie algebra is $Lie(SL(n))$, cf. [**Ka-ESDE**, 1.5]. [See [**Se-Dri**, Prop. 5] for another approach to this result.] Therefore $G_{geom,N}^{0,der} = SL(n)$, and hence we get the inclusion $SL(n) \subset G_{geom,N}$. So it suffices to show that the determinant as a character of $G_{geom,N}$ has infinite order.

Equivalently, we must show that "det"(N) in the Tannakian sense is geometrically of infinite order. The one-dimensional objects in $<N>_{arith}$ are either punctual objects $\alpha^{deg} \otimes \delta_a$ for some unitary scalar α and some $a \in k^\times$ or they are, geometrically, multiplicative translates of hypergeometric sheaves $\mathcal{H}[1]$, cf. [**Ka-ESDE**, 8.5.3]. But $\mathcal{H}[1]$ has infinite geometric order, because its successive middle convolutions with itself are again of the same form $\mathcal{H}'[1]$.

So it suffices to show that "det"(N) in the Tannakian sense is not punctual. But if it were punctual, it would have no bad characters, i.e., we would have $|\det(Frob_{E,\chi})| = 1$ for every finite extension E/k and every character χ of $E^\times$. But we have seen that $Frob_{k,\mathbb{1}}^{ss}$ is $Diag(1/\sqrt{\#k}^e, 1, 1, ..., 1)$. Hence $\det(Frob_{k,\mathbb{1}}) = 1/\sqrt{\#k}^e$ is not pure of weight zero. Therefore "det"(N) in the Tannakian sense is not punctual, and hence it is geometrically of infinite order. $\square$

Here are five explicit examples, all on $\mathbb{G}_m/k$. The first is based on the Legendre sheaf Leg introduced in chapter 15.

Theorem 17.2. *Let k have odd characteristic. For any odd integer $n \geq 1$, the object $N := \text{Sym}^n(Leg)((n+1)/2)[1]$ in $\mathcal{P}_{arith}$ is pure of weight zero, has "dimension" n, and has*

$$G_{geom,N} = G_{arith,N} = GL(n).$$

Proof. $\text{Sym}^n(Leg)(n/2)$ is the middle extension of a geometrically irreducible (because $\text{Sym}^n(std_2)$ is an irreducible representation of $SL(2)$) lisse sheaf of rank $n+1$ on $\mathbb{G}_m \setminus \{1\}$ which is pure of weight zero. So N is a geometrically irreducible object in $\mathcal{P}_{arith}$ which is pure of weight zero. Its local monodromies at 0 and 1 are both $Unip(n+1)$. Because n is odd, its local monodromy at ∞ is $\mathcal{L}_{\chi_2} \otimes Unip(n+1)$. Because the only singularity of N in $\mathbb{G}_m$ is 1, N is not geometrically isomorphic to any nontrivial multiplicative translate of itself. The Euler-Poincaré formula [**Ray**] shows that N has "dimension" n (it is tame at 0 and ∞, and has drop n at 1). The result now follows from the previous theorem. In this example, we can compute the Tannakian determinant "det"(N) explicitly. By Theorem 16.1 and Corollary 16.3, "det"(N) is geometrically isomorphic to a multiplicative translate of the shifted hypergeometric sheaf $\mathcal{H}[1]$ of type $(n+1, n+1)$ given by $\mathcal{H} = \mathcal{H}(\psi; \mathbb{1}, ..., \mathbb{1}; \chi_2, ..., \chi_2)$. [By Lemma 19.3, proven later but with no circularity, there is in fact no multiplicative translate: the unique singularity in $\mathbb{G}_m$ of both N and its determinant is at 1.] $\qquad\square$

The second example, a mild generalization of the first, is based on hypergeometric sheaves of type $(2, 2)$. We pick a character χ of $k^\times$ for a large enough finite field k. We assume χ^2 is nontrivial. We begin with the objects $\mathcal{L}_{\chi(1-x)}(1/2)[1]$ and $\mathcal{L}_{\chi(x^2/(1-x))}(1/2)[1]$ on $\mathbb{G}_m/k$.) Their middle convolution is $\mathcal{H}(2)[1]$, for $\mathcal{H} := \mathcal{H}(!, \psi; \mathbb{1}, \chi^2; \chi, \chi)$ the hypergeometric sheaf of type $(2, 2)$ on $\mathbb{G}_m/k$ as defined in [**Ka-ESDE**, 8.2]. This sheaf $\mathcal{H}$ is the middle extension of its restriction to $\mathbb{G}_m \setminus \{1\}$, where it is lisse, and pure of weight 3. Its local monodromy at 0 is $\mathbb{1} \oplus \chi^2$. Its local monodromy at 1 is $Unip(2)$, and its local monodromy at ∞ is $\mathcal{L}_\chi \otimes Unip(2)$.

Theorem 17.3. *Let $n \geq 2$ be an integer, $\mathcal{H} := \mathcal{H}(!, \psi; \mathbb{1}, \chi^2; \chi, \chi)$. If χ^2 has order $> n$, the object $N := \mathrm{Sym}^n(\mathcal{H})((3n+1)/2)[1]$ in $\mathcal{P}_{arith}$ is pure of weight zero, has "dimension" n, and has*

$$G_{geom,N} = G_{arith,N} = GL(n).$$

Proof. Because we assume that χ^2 has order $> n$, the local monodromy of $\mathrm{Sym}^n(\mathcal{H})$ at 0, which is the direct sum $\mathbb{1} \oplus \bigoplus_{r=1}^{n} \chi^{2r}$, has $\mathrm{Sym}^n(\mathcal{H})(0)^{unip} = Unip(1)$. The local monodromy of $\mathrm{Sym}^n(\mathcal{H})$ at 1 is $Unip(n+1)$, and its local monodromy at ∞ is $\mathcal{L}_{\chi^n} \otimes Unip(n+1)$. Notice that χ^n is nontrivial (otherwise $(\chi^2)^n$ is trivial). Therefore $\mathrm{Sym}^n(\mathcal{H})(\infty)^{unip} = 0$. The rest of the proof is identical to that in the Legendre case above. Just as in that case, we see that here the Tannakian determinant is geometrically isomorphic to $\mathcal{H}[1]$ for $\mathcal{H}$ the hypergeometric sheaf of type (n, n) $\mathcal{H}(\psi; \chi^2, ..., \chi^{2n}; \mathbb{1}, ..., \mathbb{1})$. $\qquad\square$

Our third example works on $\mathbb{G}_m/\mathbb{F}_p$, any prime p. For any integer $n \geq 1$, we will construct a lisse rank one sheaf $\mathcal{F}$ on $\mathbb{A}^1/\mathbb{F}_p$ which is pure of weight zero and whose Swan conductor at ∞ is the integer n. We will do this in such a way that $\mathcal{F}|\mathbb{G}_m$ is not geometrically isomorphic to any nontrivial multiplicative translate of itself. Then it is immediate from Theorem 17.1 that the object $N := (\mathcal{F}|\mathbb{G}_m)(1/2)[1]$ in $\mathcal{P}_{arith}$ is pure of weight zero, has "dimension" n, and has

$$G_{geom,N} = G_{arith,N} = GL(n).$$

Here is one such construction. Let us write $n = p^r d$ with $r \geq 0$ and with $d \geq 1$ prime to p. Suppose first that $r = 0$, i.e., $n = d$ is prime to p. Pick a nontrivial additive character ψ of $\mathbb{F}_p$. Choose a polynomial $f_d(x) \in k[x]$ which is Artin-Schreier reduced, i.e., no monomial x^e

appearing in f_d has $p|e$, and such that the set of e's such that x^e occurs in f_d generates the unit ideal in $\mathbb{Z}$. For instance we could take $f_1(x) = x$ and $f_d(x) = x^d - x$ for $d \geq 2$. Then the Artin-Schreier sheaf $\mathcal{L}_{\psi(f_d(x))}$ has Swan conductor d at ∞ [**De-ST**, 3.5.4]. Its multiplicative translate by an $a \neq 1$ in $\overline{k}^\times$ is $\mathcal{L}_{\psi(f_d(ax))}$, which is not geometrically isomorphic to $\mathcal{L}_{\psi(f_d(x))}$ because their ratio is $\mathcal{L}_{\psi(f_d(ax)-f_d(x))}$. Indeed, the difference $f_d(ax) - f_d(x)$ is Artin-Schreier reduced, and has strictly positive degree so long as $a \neq 1$; this degree is the Swan conductor at ∞ of the ratio.

Suppose next that $n = p^r d$ with $r \geq 1$ and $d \geq 1$ prime to p. Then we pick a character ψ_{r+1} of $\mathbb{Z}/p^{r+1}\mathbb{Z} = W_{r+1}(\mathbb{F}_p)$ which has order p^{r+1}. We consider the Witt vector of length $r+1$ given by $v := (f_d(x), 0, 0, .., 0) \in W_{r+1}(\mathbb{F}_p[x])$. We form the $\mathbb{Z}/p^{r+1}\mathbb{Z}$ covering of $\mathbb{A}^1$ defined by the Witt vector equation $z - F(z) = v$ in W_{r+1}; its pushout by ψ_{r+1} gives us the Artin-Schreier-Witt sheaf $\mathcal{L}_{\psi_{r+1}((f_d(x),0,0,0,..,0))}$, whose Swan conductor at ∞ is $p^r d = n$, cf. [**Bry**, Prop. 1 and Cor. of Thm. 1]. It is not geometrically isomorphic to any nontrivial multiplicative translate of itself. Indeed, its p^r'th tensor power is just $\mathcal{L}_{\psi(f_d(x))}$ for ψ the additive character of $\mathbb{F}_p$, viewed as $p^r\mathbb{Z}/p^{r+1}\mathbb{Z}$, obtained by restricting ψ_{r+1}.

In summary, then, we have the following theorem.

Theorem 17.4. *For $n = p^r d$ with $r \geq 0$ and $d \geq 1$ prime to p, form the Artin-Schreier-Witt sheaf $\mathcal{F} := \mathcal{L}_{\psi_{r+1}((f_d(x),0,0,..,0))}$. Then the object $N := (\mathcal{F}|\mathbb{G}_m)(1/2)[1]$ in $\mathcal{P}_{arith}$ is pure of weight zero, has "dimension" n, and has*

$$G_{geom,N} = G_{arith,N} = GL(n).$$

In this example, the Tannakian determinant is geometrically isomorphic to a multiplicative translate of $\mathcal{H}[1]$ for $\mathcal{H} = \mathcal{L}_{\psi(x)}$, the hypergeometric $\mathcal{H}(\psi; \mathbb{1}; \emptyset)$ of type $(1, 0)$.

Our fourth example is this.

Theorem 17.5. *Take a polynomial $f[x] = \sum_{i=0}^n A_i x^i$ in $k[x]$ of degree $n \geq 2$ with all distinct roots in $\overline{k}$. Suppose that $f(0) \neq 0$, and that $\gcd\{i | A_i \neq 0\} = 1$. Then for any character χ of $k^\times$ such that χ^n is nontrivial, the object $N := \mathcal{L}_{\chi(f)}(1/2)[1]$ in $\mathcal{P}_{arith}$ is pure of weight zero, has "dimension" n, and has*

$$G_{geom,N} = G_{arith,N} = GL(n).$$

Proof. Here the local monodromy at 0 is $Unip(1)$, and the local monodromy at ∞ is $\mathcal{L}_{\chi^n}$. So the assertion is an immediate application of Theorem 17.1, once we show that N is isomorphic to no nontrivial multiplicative translate of itself. To see this, we argue as follows. The set

S of zeroes of f is the set of finite singularities of N, so is an intrinsic invariant of the geometric isomorphism class of N. So it suffices to show that if $\alpha \in \overline{k}^{\times}$ satisfies $\alpha S = S$, then $\alpha = 1$. The condition that $\alpha S = S$ is the condition that $f(\alpha x) = \alpha^n f(x)$. Equating coefficients, we get the equations

$$\alpha^i A_i = \alpha^n A_i$$

for every i. Taking $i = 0$, and remembering that $A_0 \neq 0$ by hypothesis, we find that $\alpha^n = 1$. Then our equations read

$$\alpha^i A_i = A_i$$

for every i. Using the fact that $gcd\{i|A_i \neq 0\} = 1$, we infer that $\alpha = 1$. $\square$

In this example, the Tannakian determinant is geometrically isomorphic to $\mathcal{H}[1]$ for $\mathcal{H} = \mathcal{H}(\psi; \mathbb{1}; \chi^n) \cong \mathcal{L}_{\chi^n(1-x)}$.

To end this chapter, we give a fifth example, valid in any odd characteristic p.

Theorem 17.6. *Let k be a finite field of odd characteristic p. Take a polynomial $f(x) = \sum_{i=0}^n A_i x^i$ in $k[x]$ of prime-to-p degree $n \geq 2$ with all distinct roots in $\overline{k}$. Suppose that f is "weakly supermorse" [**Ka-ACT,** 5.5.2], i.e., its derivative $f'(x)$ has $n-1$ distinct zeroes (the critical points) α_i in $\overline{k}$, and the $n-1$ values $f(\alpha_i)$ (the critical values) are all distinct. Denote by S the set of critical values. Suppose that S is not equal to any nontrivial multiplicative translate of itself. Form the middle extension sheaf*

$$\mathcal{F} := f_\star \overline{\mathbb{Q}_\ell} / \overline{\mathbb{Q}_\ell} \mid \mathbb{G}_m.$$

Then the object $N := \mathcal{F}(1/2)[1] \in \mathcal{P}_{arith}$ is pure of weight zero, has "dimension" $n-1$, and has

$$G_{geom,N} = G_{arith,N} = GL(n-1).$$

Proof. Because f is weakly supermorse, $\mathcal{F}$ is irreducible, of generic rank $n-1$, with geometric monodromy group the symmetric group S_n in its deleted permutation representation, cf. [**Ka-ESDE,** 7.10.2.3 and its proof]. Because f has n distinct zeroes, 0 is not a critical value. So $S \subset \mathbb{G}_m$, and at each point of S the local monodromy is a reflection, necessarily tame, as $p \neq 2$. At ∞, the local monodromy is tame, the direct sum $\oplus_{\chi|\chi^n=1,\chi\neq\mathbb{1}} \mathcal{L}_\chi$. Thus $\mathcal{F}$ is tame at ∞. Therefore the "dimension" of N is the sum of the drops of $\mathcal{F}$, each one, at the $n-1$ points of S. As $\mathcal{F}$ is not lisse at $n-1 \geq 1$ points of $\mathbb{G}_m$, $\mathcal{F}$ is not geometrically isomorphic to any Kummer sheaf $\mathcal{L}_\rho$. Thus $N \in \mathcal{P}_{arith}$. It is pure of weight zero because $\mathcal{F}$ is a middle extension which on a dense open set is pure of weight zero. From the hypothesis that the set S of singularities of N is not equal to any nontrivial multiplicative translate of itself, it follows that N is not geometrically isomorphic to any nontrivial multiplicative translate of itself. Thus N is geometrically Lie-irreducible.

It remains to show that N has $G_{geom} = GL(n-1)$. For this, it is equivalent to show that for some ρ, $N \otimes \mathcal{L}_\rho$ has $G_{geom} = GL(n-1)$ (simply because $M \mapsto M \otimes \mathcal{L}_\rho$ is a Tannakian automorphism of $\mathcal{P}_{geom}$).

We choose for ρ any nontrivial character of order dividing n. For such a choice of ρ, the sheaf $\mathcal{F}_\rho := \mathcal{F} \otimes \mathcal{L}_\rho$ has $\mathcal{F}_\rho(0)^{unip} = 0$ (because $\mathcal{F}$ was lisse at 0), and $\mathcal{F}_\rho(\infty)^{unip} = Unip(1)$ (because the local monodromy of $\mathcal{F}$ at ∞ was $\oplus_{\chi|\chi^n=\mathbb{1},\chi\neq\mathbb{1}}\mathcal{L}_\chi$). The result now follows from Theorem 17.1, using the "other" geometric isomorphism of $\mathbb{G}_m$ with itself (which interchanges 0 and ∞).   □

In this case, the Tannakian determinant of the original N is geometrically isomorphic to some multiplicative translate of $\mathcal{H}[1]$ for $\mathcal{H}$ the hypergeometric sheaf of type $(n-1, n-1)$ $\mathcal{H}(\psi; \mathbb{1}, ..., \mathbb{1}; \chi, \chi^2, ..., \chi^{n-1})$ for any character χ of full order n.

Remark 17.7. For any polynomial f, the trace function of the sheaf $\mathcal{F} := f_\star \overline{\mathbb{Q}_\ell}/\overline{\mathbb{Q}_\ell}$ is the counting function given, for any finite extension E/k and any $a \in E^\times$, by

$$\mathrm{Trace}(Frob_{E,a}|\mathcal{F}) = \#\{x \in E | f(x) = a\} - 1.$$

So for any nontrivial character χ of $E^\times$, we have

$$\mathrm{Trace}(Frob_{E,\chi}|\omega(N)) = (-1/\sqrt{\#E}) \sum_{x \in E} \chi(f(x)).$$

Thus for f satisfying the hypotheses of Theorem 17.6, we are saying, in particular, that these sums, as χ varies over characters of $E^\times$ with $\chi^n \neq \mathbb{1}$, are approximately distributed like the traces of random elements of the unitary group $U(n-1)$, the approximation getting better and better as $\#E$ grows.

How restrictive are the hypotheses imposed on the polynomial f? One knows that given **any** polyonomial $f(x) \in k[x]$ of prime-to-p degree $n \geq 2$ such that $f''(x)$ is nonzero, then for all but finitely many values of $a \in \overline{k}$, the polynomial $f(x) + ax$ is weakly supermorse, cf. [**Ka-ACT**, 5.15]. For example, if $n(n-1)$ is prime to p, then $x^n - nx$ is weakly supermorse, with μ_{n-1} as the critical points, and $(1-n)\mu_{n-1}$ as the set S of critical values. In this example, the set S is equal to its multiplicative translate by any element of μ_{n-1}. Nonetheless, we have the following lemma, which shows that by adding nearly any constant to a weakly supermorse f, the set of its critical values will then be equal to no nontrivial multiplicative translate of itself.

Lemma 17.8. *Let k be a field of characteristic p. Take a weakly super-morse polynomial $f[x] = \sum_{i=0}^{n} A_i x^i$ in $k[x]$ of prime-to-p degree $n \geq 2$, with set S of critical values. For each $\lambda \in \overline{k}$, consider the weakly super-morse polynomial $f_\lambda(x) := f(x) - \lambda$, with set $S_\lambda := -\lambda + S$ of critical values. Denote by $F_S(x) := \prod_{s \in S}(x - s)$ the polynomial of degree $n - 1$ whose roots are the critical values of f. For any $\lambda \in \overline{k}$ such that both $F_S(\lambda)$ and its derivative $F'_S(\lambda)$ are nonzero, the set S_λ is equal to no nontrivial multiplicative translate of itself. In particular, there are at most $(n - 1)(n - 2)$ values of $\lambda \in \overline{k}$ for which S_λ is equal to some nontrivial multiplicative translate of itself.*

This results from the following elementary lemma.

Lemma 17.9. *Let k be a field, $S \subset k$ a finite subset consisting of $d \geq 1$ elements. Denote by $F_S(x) := \prod_{s \in S}(x - s)$ the polynomial of degree d whose roots are the points of S. For any $\lambda \in \overline{k}$ such that both $F_S(\lambda)$ and its derivative $F'_S(\lambda)$ are nonzero, the set $S_\lambda := -\lambda + S$ is equal to no nontrivial multiplicative translate of itself.*

Proof. The polynomial $F_{S_\lambda}(x)$ is just $F_S(x + \lambda)$, hence we have the Taylor expansion

$$F_{S_\lambda}(x) = F_S(\lambda) + F'_S(\lambda)x \mod x^2.$$

Suppose $F_S(\lambda)F'_S(\lambda) \neq 0$. If S_λ is equal to its multiplicative translate by some nonzero α, then $F_{S_\lambda}(\alpha x)$ has the same roots as $F_{S_\lambda}(x)$. Comparing highest degree terms, we get

$$F_{S_\lambda}(\alpha x) = \alpha^d F_{S_\lambda}(x).$$

Comparing the *nonzero* (by our choice of λ) constant and linear terms of their Taylor expansions, we get $\alpha^d = 1$ and $\alpha^{d-1} = 1$, and hence $\alpha = 1$. □

Corollary 17.10. *Let k be a finite field of odd characteristic p. Let $n \geq 2$, and assume that $n(n-1)$ is prime to p. Consider the polynomial*

$$f(x) = \frac{(x^n - nx)}{(1 - n)}.$$

Then for any $a \in k^\times$ with $a^{n-1} \neq 1$, the polynomial $f(x) - a$ satisfies all the hypotheses of Theorem 17.6.

Proof. Here the set S of critical values of $f(x)$ is μ_{n-1}, so $F_S(x) = x^{n-1} - 1$, and $F'_S(x) = (n - 1)x^{n-2}$. □

There are some special cases where no adjustment of the constant term is necessary. Here is one such.

Lemma 17.11. *Let $k = \mathbb{F}_q$ be a finite field of odd characteristic. For any $a \in \mathbb{F}_q^\times$, the polynomial $f(x) = x^{q+1} - x^2/2 + ax$ satisfies all the hypotheses of Theorem 17.6.*

Proof. We have $f'(x) = x^q - x + a$, so the critical values are the elements $\alpha \in \overline{k}$ with $\alpha^q = \alpha - a$. No such α lies in $\mathbb{F}_q$, since $a \neq 0$. The value of f at such an α is

$$f(\alpha) = \alpha(\alpha^q) - \alpha^2/2 + a\alpha = \alpha(\alpha - a) - \alpha^2/2 + a\alpha = \alpha^2/2.$$

If we fix one critical point α, any other is $\alpha + b$, for some $b \in \mathbb{F}_q$. If their critical values coincide, i.e., if $(\alpha + b)^2/2 = \alpha^2/2$, then $\alpha b + b^2/2 = 0$; if $b \neq 0$, this implies that $\alpha = -b/2$, contradicting the fact that α does not lie in $\mathbb{F}_q$. So the set S of critical values consists of q distinct, nonzero elements. So the polynomial $F_S(x)$ has degree q, and a nonzero constant term. If S is a multiplicative translate of itself, say by $\gamma \in \overline{k}^\times$, then we get $F_S(\gamma x) = \gamma^q F_S(x)$, and comparing the nonzero constant terms we get $\gamma^q = 1$, and hence $\gamma = 1$.   □

CHAPTER 18

Symplectic Examples

We work on either the split or the nonsplit form. We begin with a lisse sheaf $\mathcal{F}$ on a dense open set $j : U \subset G$ which is geometrically irreducible, pure of weight zero, and not geometrically isomorphic to (the restriction to U of) any Kummer sheaf $\mathcal{L}_\chi$. We denote by $\mathcal{G} := j_\star \mathcal{F}$ its middle extension to G. Then the object $N := \mathcal{G}(1/2)[1] \in \mathcal{P}_{arith}$ is pure of weight zero and geometrically irreducible.

Theorem 18.1. *Suppose that N is not geometrically isomorphic to any nontrivial multiplicative translate of itself, and that N is symplectically self-dual. Suppose further that for either of the two possible geometric isomorphisms $G/\overline{k} \cong \mathbb{G}_m/\overline{k}$, both $\mathcal{F}(0)^{unip}$ and $\mathcal{F}(\infty)^{unip}$ are a single Jordan block $Unip(e)$ of the same size $e \geq 1$. For $n := \dim(\omega(N))$ we have*

$$G_{geom,N} = G_{arith,N} = Sp(n).$$

Proof. We have a priori inclusions

$$G_{geom,N} \subset G_{arith,N} \subset Sp(n),$$

so it suffices to prove that $G_{geom,N} = Sp(n)$. So we may extend scalars if necessary from k to its quadratic extension k_2, and reduce to the case where G is $\mathbb{G}_m$. The hypothesis that N is not geometrically isomorphic to any nontrivial multiplicative translate of itself insures that N is geometrically Lie-irreducible, i.e., that G^0_{geom} is a connected irreducible subgroup of $Sp(n)$. Thus G^0_{geom} is semisimple: any connected irreducible subgroup of $Sp(n)$ (indeed of $SL(n)$) is semisimple (it is reductive, because irreducible, and its center, necessarily consisting entirely of scalars by irreducibility, is finite).

The local monodromy of N at both 0 and ∞ is $Unip(e)$. Therefore the semisimplification of $Frob_{k,1}$ gives us a Frobenius torus [**Se-Let**] $Diag(x, 1/x, 1, ..., 1)$ in $G_{arith,N}$. This torus normalizes the connected semisimple group $G^0_{geom,N}$, and this group is an irreducible (in the given n-dimensional representation) subgroup of $Sp(n)$. Take an element $x_0 \in \overline{\mathbb{Q}_\ell}^\times$ which is not a root of unity. Then the element $diag(x_0, 1/x_0, 1, ..., 1)$ normalizes $G^0_{geom,N}$. But a fixed power of any automorphism of a connected semisimple group is inner. So for some integer $d \geq 1$, the element $diag(x_0^d, 1/x_0^d, 1, ..., 1)$ induces an inner auto-morphism of $G^0_{geom,N}$, conjugation by some element $\gamma \in G^0_{geom,N}$. Then $\gamma^{-1} diag(x_0^d, 1/x_0^d, 1, ..., 1)$ commutes with every element of $G^0_{geom,N}$, so is a scalar, say $t \in \overline{\mathbb{Q}_\ell}^\times$. Thus we find

$$diag(x_0^d, 1/x_0^d, 1, ..., 1) = t\gamma,$$

equality inside $\mathbb{G}_m G^0_{geom,N} \subset GL(n)$. Both $diag(x_0^d, 1/x_0^d, 1, ..., 1)$ and γ have determinant one, so comparing determinants we see that $t^n = 1$, and hence

$$diag(x_0^{nd}, 1/x_0^{nd}, 1, ..., 1) = \gamma^n$$

lies in $G^0_{geom,N}$. As x_0 is not a root of unity, the entire torus $Diag(x, 1/x, 1, ..$ lies in $G^0_{geom,N}$. By a beautiful result of Kostant and Zarhin [**Ka-ESDE,** 1.2], the only irreducible connected semisimple subgroups of $SL(n)$ which contain the torus $Diag(x, 1/x, 1, ..., 1)$ are $SL(n)$, $SO(n)$, and, when n is even, $Sp(n)$. Since we have an a priori inclusion $G^0_{geom,N} \subset Sp(n)$, we must have $G^0_{geom,N} = Sp(n)$. From the a priori inclusions $G^0_{geom,N} \subset G_{geom,N} \subset G_{arith,N} \subset Sp(n)$, we get the asserted conclusion. □

Here is a generalization of this last result.

Theorem 18.2. *Suppose N satisfies all the hypotheses of the theorem above. Suppose further that*

> *At either 0 or at ∞ or at both, the entire tame part of the local monodromy is $Unip(e)$, i.e., local monodromy there is the direct sum of $Unip(e)$ and of something totally wild.*

Then for any $s \geq 2$ distinct characters $\chi_1, ..., \chi_s$ of $G(k)$, the objects $N_i := N \otimes \mathcal{L}_{\chi_i}$ have

$$G_{geom,\oplus_{i=1}^s N_i} = G_{arith,\oplus_{i=1}^s N_i} = \prod_{i=1}^s Sp(n).$$

Proof. This is an immediate application of Theorem 13.1. As already noted, the operation $M \mapsto M \otimes \mathcal{L}_{\chi_i}$ is a Tannakian isomorphism from $<N>_{arith}$ to $<N \otimes \mathcal{L}_{\chi_i}>_{arith}$, so each N_i has $G_{geom,N_i} = G_{arith,N_i} = Sp(n)$. For $i \neq j$, N_i and N_j have nonisomorphic tame parts of local monodromy at either 0 or at ∞ or at both, so they cannot be geo-metrically isomorphic. For the same reason, N_i is not geometrically isomorphic to $[x \mapsto -x]^\star N_j$; indeed N_j and $[x \mapsto -x]^\star N_j$ have isomor-phic tame parts of local monodromy at both 0 and ∞. [More generally, tame representations of either $I(0)$ or of $I(\infty)$ are geometrically iso-morphic to all their multiplicative translates.] □

We now give some concrete examples. The first example is based on the Legendre sheaf Leg introduced in Chapter 15.

Theorem 18.3. *Let k have odd characteristic. For any even integer $n \geq 2$, the object $N := \operatorname{Sym}^n(Leg)((n+1)/2)[1]$ in $\mathcal{P}_{arith}$ is pure of weight zero, of "dimension" n, and has*

$$G_{geom,N} = G_{arith,N} = Sp(n).$$

Proof. The lisse sheaf $Leg(1/2)|\mathbb{A}^1 \setminus \{0,1\}$ has $SL(2)$ as its geometric (and its arithmetic) monodromy group, so every symmetric power Sym^n of it is geometrically irreducible and self-dual, orthogonally if n is even and symplectically if n is odd. Under multiplicative inversion, we have

$$[\lambda \mapsto 1/\lambda]^* Leg \cong Leg \otimes \mathcal{L}_{\chi_2(\lambda)}.$$

Thus each even symmetric power of $Leg(1/2)|\mathbb{A}^1 \setminus \{0,1\}$ is both orthogonally self-dual and invariant under multiplicative inversion. So the object N in $\mathcal{P}_{arith}$ is pure of weight zero, geometrically irreducible, and self-dual. Moreover it is geometrically Lie-irreducible, because the point $1 \in \mathbb{G}_m(\overline{k})$ is its unique singularity. Its "dimension" is n; indeed it is everywhere tame, and at its unique singularity $1 \in \mathbb{G}_m(\overline{k})$ its local monodromy is $Unip(n+1)$, so its drop is n. Let us admit for the moment that the autoduality of N is symplectic. Then we argue as follows. The local monodromy of N at both 0 and ∞ is $Unip(n+1)$. Theorem 18.1 then gives the asserted conclusion.

It remains to show that N is symplectically self-dual. Because we know already that N is either symplectically or orthogonally self-dual, we may make a finite extension of the ground field to determine which. So replacing k by k_2 if necessary, we reduce to the case when k contains the fourth roots of unity. We will apply Theorem 11.2. Write the even integer n as $n = 2d$. We will show that the lisse sheaf $\operatorname{Sym}^{2d}(Leg)(d)|\mathbb{A}^1 \setminus \{0,1\}$ is the pullback, by $\lambda \mapsto \lambda + 1/\lambda$, of a lisse sheaf $\mathcal{F}_{2d+1}$ on $\mathbb{A}^1 \setminus \{2,-2\}$ which is orthogonally self-dual, and whose $G_{geom,\mathcal{F}_{2d+1}}$ is the image of $\mu_4 SL(2)$ in $Sym^{2d}(std_2)$. [In this representation, $SL(2)$ has image an irreducible subgroup of $SO(2d+1)$, while μ_4 acts trivially if d is even, and with scalar image ± 1 if d is odd.] Then by Theorem 11.2 the autoduality on N is symplectic.

For the group $SL(2)$, we recover $\operatorname{Sym}^{2d}(std_2)$ as the highest dimensional constituent of $\operatorname{Sym}^d(\operatorname{Sym}^2(std_2))$. Indeed, by Hermite's identity, we have $\operatorname{Sym}^d(\operatorname{Sym}^2(std_2)) \cong \operatorname{Sym}^2(\operatorname{Sym}^d(std_2))$, and as already noted (in the paragraph preceding Theorem 11.2), we have

$$\mathrm{Sym}^2(\mathrm{Sym}^d(std_2)) = \bigoplus_{r=0}^{[d/2]} \mathrm{Sym}^{2d-4r}(std_2) \otimes det^{\otimes 2r}.$$

Now think of $\mathrm{Sym}^2(std_2)$ as std_3 for $SO(3)$. This group $SO(3)$ has a unique irreducible representation V_{2r+1} of each odd dimension $2r+1$, namely $\mathrm{Sym}^{2r}(std_2)$, when we think of $SO(3)$ as $SL(2)/\pm 1$. The above identity then is the statement that as representations of $SO(3)$, we have

$$\mathrm{Sym}^d(std_3) \cong \bigoplus_{r=0}^{[d/2]} V_{2d+1-4r}.$$

Consequently, we have

$$V_{2d+1} \cong \mathrm{Sym}^d(std_3)/\mathrm{Sym}^{d-2}(std_3).$$

[In other words, V_{2d+1} as representation of $SO(3)$ is the space of spherical harmonics of degree d on S^2.] Applying this to $Leg(1/2)$ on $\mathbb{A}^1 \setminus \{0,1\}$, we get

$$Sym^{2d}(Leg)(d) \cong \mathrm{Sym}^d(Sym^2(Leg)(1))/\mathrm{Sym}^{d-2}(Sym^2(Leg)(1)).$$

This reduces us to the case $n = 2$, i.e., to the problem of showing that $\mathrm{Sym}^2(Leg)(1)|\mathbb{A}^1 \setminus \{0,1\}$ is arithmetically isomorphic to the pullback, by $\lambda \mapsto \lambda + 1/\lambda$, of a lisse rank three sheaf $\mathcal{F}_3$ on $\mathbb{A}^1 \setminus \{2,-2\}$ which is orthogonally self-dual, with $G_{geom,\mathcal{F}_3}$ the group $O(3) = \pm SO(3)$, viewed as the image of $\mu_4 SL(2)$ in $\mathrm{Sym}^2(std_2)$. To do this, we may replace Leg by its quadratic twist $TwLeg := Leg \otimes \mathcal{L}_{\chi2(1-\lambda)}$, because for even n, Leg and $TwLeg$ have the same Sym^n.

To construct $\mathcal{F}_3$, we will first exhibit a lisse, rank two sheaf $\mathcal{H}$ on $\mathbb{A}^1 \setminus \{2,-2\}$ such that $TwLeg$ is geometrically the pullback of $\mathcal{H}$ by $\lambda \mapsto \lambda + 1/\lambda$. This sheaf $\mathcal{H}$ is not self-dual, but we will show that $\mathrm{Sym}^2(\mathcal{H})$ is self-dual, with $G_{geom,\mathrm{Sym}^2(\mathcal{H})}$ the group $O(3)$. Then we will take $\mathcal{F}_3$ to be $\mathrm{Sym}^2(\mathcal{H})$, or possibly its constant field quadratic twist $(-1)^{deg} \otimes \mathrm{Sym}^2(\mathcal{H})$.

For $\mathcal{H}$, we start with the hypergeometric sheaf $\mathcal{H}(\psi; \chi_4, \chi_4; \mathbb{1}, \mathbb{1})(3/2)$ of type $(2,2)$, with χ_4 a character of $k^\times$ of order 4. This sheaf is pure of weight zero, its local monodromy at 0 is $\mathcal{L}_{\chi_4} \otimes Unip(2)$, its local monodromy at 1 is $\mathbb{1} \oplus \mathcal{L}_{\chi_2}$, and its local monodromy at ∞ is $Unip(2)$. Then $[x \mapsto (2-x)/4]^\star \mathcal{H}(\psi; \chi_4, \chi_4; \mathbb{1}, \mathbb{1})(3/2)$ is the desired $\mathcal{H}$. Its local monodromy at ∞ is $Unip(2)$, its local monodromy at 2 is $\mathcal{L}_{\chi_4} \otimes Unip(2)$, and its local monodromy at -2 is $\mathbb{1} \oplus \mathcal{L}_{\chi_2}$.

The pullback of this $\mathcal{H}$ by $\lambda \mapsto \lambda + 1/\lambda$ has the same local monodromies as $TwLeg$, namely $Unip(2)$ at 0 and ∞, and $\mathcal{L}_{\chi_2} \otimes Unip(2)$

at 1. The pullback is lisse at -1. By rigidity, the pullback is geometrically isomorphic to $TwLeg$. Therefore $\text{Sym}^2(TwLeg)(1)|\mathbb{A}^1 \setminus \{0, 1\}$ is geometrically isomorphic to the pullback by $\lambda \mapsto \lambda + 1/\lambda$ of

$$\mathcal{F}_3 := \text{Sym}^2(\mathcal{H}) = [x \mapsto (2 - x)/4]^*\text{Sym}^2(\mathcal{H}(\psi; \chi_4, \chi_4; \mathbb{1}, \mathbb{1}))(3).$$

Again by rigidity, $\text{Sym}^2(\mathcal{H}(\psi; \chi_4, \chi_4; \mathbb{1}, \mathbb{1}))(3)$ is geometrically isomorphic to the hypergeometric sheaf $\mathcal{H}(\chi_2, \chi_2, \chi_2; \mathbb{1}, \mathbb{1}, \mathbb{1})(5/2)$ of type $(3, 3)$. This sheaf is orthogonally self-dual. We claim that its G_{geom} is $O(3)$. Its local monodromy at ∞ is $Unip(3)$, so $I(\infty)$, and a fortiori G^0_{geom}, act indecomposably. Therefore G_{geom} acts irreducibly (because by purity it acts semisimply). So G^0_{geom} is a connected irreducible subgroup of $SO(3)$, and the only such is $SO(3)$ itself. On the other hand, $\det(\mathcal{H}(\chi_2, \chi_2, \chi_2; \mathbb{1}, \mathbb{1}, \mathbb{1})(5/2))$ is geometrically $\mathcal{L}_{\chi_2(x(1-x))}$, so we have $G_{geom} = O(3)$, as asserted.

So our situation is this. We have a lisse sheaf $\mathcal{F}_3$ which is pure of weight zero, lisse of rank three, orthogonally self-dual with $G_{geom,\mathcal{F}_3} = O(3)$. Its pullback by $\lambda \mapsto \lambda + 1/\lambda$, call it $\mathcal{K}$, is geometrically isomorphic to $\text{Sym}^2(TwLeg)(1)$. Since both $\text{Sym}^2(TwLeg)(1)$ and the pullback $\mathcal{K}$ are geometrically irreducible and arithmetically orthogonal, the space

$$Hom_{geom}(\mathcal{K}, \text{Sym}^2(TwLeg)(1))$$

is a one-dimensional orthogonal $Gal(\bar{k}/k)$-representation, some α^{deg}, with $\alpha = \pm 1$. We have an arithmetic isomorphism

$$\mathcal{K} \otimes Hom_{geom}(\mathcal{K}, \text{Sym}^2(TwLeg)(1)) \cong \text{Sym}^2(TwLeg)(1),$$

i.e., an arithmetic isomorphism $\mathcal{K} \otimes \alpha^{deg} \cong \text{Sym}^2(TwLeg)(1)$. Replacing $\mathcal{F}_3$ by $\alpha^{deg} \otimes \mathcal{F}_3$ if necessary, i.e., if $\alpha = -1$, we get the required $\mathcal{F}_3$. $\qquad \square$

Applying Theorem 18.2 to this N, we get the following result.

Theorem 18.4. *For $n \geq 2$ even, $N := \text{Sym}^n(Leg)((n + 1)/2)[1]$, and for any $s \geq 2$ distinct characters $\chi_1, ..., \chi_s$ of $G(k)$, the objects $N_i := N \otimes \mathcal{L}_{\chi_i}$ have*

$$G_{geom,\oplus_{i=1}^s N_i} = G_{arith,\oplus_{i=1}^s N_i} = \prod_{i=1}^s Sp(n).$$

We now analyze the **odd** symmetric powers of the twisted Legendre sheaf $TwLeg$.

Theorem 18.5. *Let k have odd characteristic and contain the fourth roots of unity. For any odd integer $n \geq 1$, the object*

$$N := \text{Sym}^n(TwLeg)((n + 1)/2)[1]$$

in $\mathcal{P}_{arith}$ is pure of weight zero, of "dimension" $n+1$, and has

$$G_{geom,N} = G_{arith,N} = Sp(n+1).$$

Proof. Because k contains the fourth roots of unity, the sheaf $TwLeg$ and hence the object N is isomorphic to its pullback by multiplicative inversion. Just as in the start of the proof of Theorem 18.3, we see that N is pure of weight zero, geometrically Lie-irreducible, and self-dual. Its local monodromy at both 0 and ∞ is $Unip(n+1)$. Its local monodromy at 1 is $\mathcal{L}_{\chi_2} \otimes Unip(n+1)$, so its "dimension" is $drop_1 = n+1$. Exactly as in Theorem 18.3, it suffices to show that N is symplectically self-dual.

As already shown in the proof of Theorem 18.3, $TwLeg$ is geometrically the pullback by $\pi : \lambda \mapsto \lambda + 1/\lambda$ of the sheaf

$$\mathcal{H} := [x \mapsto (2-x)/4]^\star \mathcal{H}(\psi; \chi_4, \chi_4; \mathbb{1}, \mathbb{1})(3/2).$$

From the known local monodromies of $\mathcal{H}(\psi; \chi_4, \chi_4; \mathbb{1}, \mathbb{1})$, we see that $\det(\mathcal{H}(\psi; \chi_4, \chi_4; \mathbb{1}, \mathbb{1}))$ is geometrically isomorphic to $\mathcal{L}_{\chi_2(x(1-x))}$. Hence $\det(\mathcal{H})$ is geometrically isomorphic to $[x \mapsto (2-x)/4]^\star \mathcal{L}_{\chi_2(x(1-x))} \cong \mathcal{L}_{\chi_2(x^2-4)}$. Thus for some unitary scalar α, $\det(\mathcal{H})$ is arithmetically isomorphic to $\alpha^{deg} \otimes \mathcal{L}_{\chi_2(x^2-4)}$.

Choose a square root β of $1/\alpha$, and put $\mathcal{H}_1 := \beta^{deg} \otimes \mathcal{H}$. This object $\mathcal{H}_1$ has determinant $\mathcal{L}_{\chi_2(x^2-4)}$ arithmetically, its pullback by $\pi : \lambda \mapsto \lambda + 1/\lambda$ is geometrically isomorphic to $TwLeg$, and its determinant is arithmetically trivial:

$$\det(\pi^\star \mathcal{H}_1) = \pi^\star \det(\mathcal{H}_1) = \pi^\star \mathcal{L}_{\chi_2(x^2-4)}$$

$$= \mathcal{L}_{\chi_2((\lambda+1/\lambda)^2-4)} = \mathcal{L}_{\chi_2((\lambda-1/\lambda)^2)} \cong \overline{\mathbb{Q}_\ell}.$$

Thus both $\pi^\star \mathcal{H}_1$ and $TwLeg(1/2)$ are arithmetically symplectically self-dual (being of rank two and trivial determinant), geometrically irreducible, and geometrically isomorphic. So for some choice of $\gamma = \pm 1$, we have an arithmetic isomorphism

$$\pi^\star(\gamma^{deg} \mathcal{H}_1) \cong TwLeg(1/2).$$

Replacing $\mathcal{H}_1$ by $\gamma^{deg} \mathcal{H}_1$ if needed, we have an $\mathcal{H}_1$ with arithmetic determinant $\mathcal{L}_{\chi_2(x^2-4)}$, geometric monodromy $\mu_4 SL(2)$, and an arithmetic isomorphism $\pi^\star \mathcal{H}_1 \cong TwLeg(1/2)$.

Now fix an odd integer $n = 2d - 1 \geq 1$. Define the lisse sheaf $\mathcal{F}$ on $\mathbb{G}_m \setminus \{\pm 2\}$ by

$$\mathcal{F} := \mathrm{Sym}^{2d-1}(\mathcal{H}_1),$$

and the object $N \in \mathcal{P}_{arith}$ by

$$N := j_\star \pi^\star \mathcal{F}(1/2)[1].$$

This N is just $\mathrm{Sym}^n(TwLeg)((n+1)/2)[1]$. The situation is reminiscent of that considered in Theorems 11.1 and 11.2, except that $\mathcal{F}$ is **not** geometrically self-dual. Indeed, the geometric monodromy group of $\mathcal{F}$ is $\mu_4 SL(2)$, viewed in the representation $Sym^{2d-1}(std_2)$, a subgroup of $GL(2d)$ which contains the scalars μ_4, hence lies in neither the orthogonal nor the symplectic group. Nonetheless, we will use the calculational method of the proof of Theorem 11.1 to determine the sign ϵ_N of the autoduality of N. In the notations of that theorem, where the "x" above becomes "s," the sums whose large $\#E$ limit is $-\epsilon_N$ are

$$(1/\#E) \sum_{s \in U(E)} \mathrm{Trace}(Frob_{E_2,s}|\mathcal{F})(1 - \chi_{2,E}(s^2 - 4)).$$

We next observe that because $\mathcal{F}$ is pure of weight zero, geometrically irreducible and **not** geometrically self-dual, we have the estimate

$$(1/\#E) \sum_{s \in U(E)} \mathrm{Trace}(Frob_{E_2,s}|\mathcal{F}) = O(1/\sqrt{\#E}),$$

cf. [**Ka-MMP**, 1.9.6, assertion 1)]. Indeed, both the sheaves $Sym^2(\mathcal{F})$ and $\Lambda^2(\mathcal{F})$ have vanishing H_c^2 on $U \otimes_k \overline{k}$, and their H_c^1's are mixed of weight ≤ 1, so the estimate follows from the Lefschetz Trace formula [**Gr-Rat**] and the linear algebra identity

$$\mathrm{Trace}(Frob_{E_2,s}|\mathcal{F}) = \mathrm{Trace}(Frob_{E,s}|Sym^2(\mathcal{F})) - \mathrm{Trace}(Frob_{E,s}|\Lambda^2(\mathcal{F})).$$

So we are reduced to showing that the sums

$$(1/\#E) \sum_{s \in U(E)} \mathrm{Trace}(Frob_{E_2,s}|\mathcal{F})\chi_{2,E}(s^2 - 4)$$

are approximately -1. To see this, we make use of the linear algebra identity above. So it suffices to prove that the Sym^2 term

$$(1/\#E) \sum_{s \in U(E)} \mathrm{Trace}(Frob_{E,s}|Sym^2(\mathcal{F}))\chi_{2,E}(s^2 - 4)$$

$$= (1/\#E) \sum_{s \in U(E)} \mathrm{Trace}(Frob_{E,s}|Sym^2(\mathcal{F}) \otimes \mathcal{L}_{\chi_2(s^2-4)}) = O(1/\sqrt{\#E})$$

and that the Λ^2 term

$$(1/\#E) \sum_{s \in U(E)} \mathrm{Trace}(Frob_{E,s}|\Lambda^2(\mathcal{F}))\chi_{2,E}(s^2 - 4)$$

$$= (1/\#E) \sum_{s \in U(E)} \mathrm{Trace}(Frob_{E,s}|\Lambda^2(\mathcal{F}) \otimes \mathcal{L}_{\chi_2(s^2-4)})$$

is approximately 1.

We now use the fact that $\mathcal{F} = \mathrm{Sym}^{2d-1}(\mathcal{H}_1)$, with $\mathcal{H}_1$ pure of weight zero, with geometric monodromy group $\mu_4 SL(2)$ and with arithmetic determinant $\mathcal{L}_{\chi_2(s^2-4)}$. Recall the identities for $GL(2)$-representations

$$\mathrm{Sym}^2(\mathrm{Sym}^{2d-1}(std_2)) = \bigoplus_{r=0}^{[(2d-1)/2]} \mathrm{Sym}^{4d-2-4r}(std_2) \otimes det^{\otimes 2r}$$

and

$$\Lambda^2(\mathrm{Sym}^{2d-1}(std_2)) = \bigoplus_{r=0}^{[(2d-2)/2]} \mathrm{Sym}^{4d-4-4r}(std_2) \otimes det^{\otimes 2r+1}.$$

Applying these to $\mathcal{H}_1$, we see every geometrically irreducible constituent of $\mathrm{Sym}^2(\mathcal{F})$ has rank ≥ 3, which gives the asserted $O(1/\sqrt{\#E})$ estimate for the Sym^2 sum. We also see that $\Lambda^2(\mathcal{F})$ is the direct sum of $\mathcal{L}_{\chi_2(s^2-4)}$ with a complement, each of whose geometrically irreducible constituents has dimension ≥ 5. So this complement gives a sum which is $O(1/\sqrt{\#E})$. The remaining term, the $\mathcal{L}_{\chi_2(s^2-4)}$ sum, is

$$= (1/\#E) \sum_{s\in E^\times\backslash\{\pm 2\}} \mathrm{Trace}(Frob_{E,s}|\mathcal{L}_{\chi_2(s^2-4)}\otimes\mathcal{L}_{\chi_2(s^2-4)}) = (\#E-3)/\#E,$$

as required. $\square$

Exactly as for even symmetric powers, we can apply Theorem 18.2 to this N.

Theorem 18.6. *For $n \geq 1$ odd, $N := \mathrm{Sym}^n(TwLeg)((n + 1)/2)[1]$, and for any $s \geq 2$ distinct characters $\chi_1, ..., \chi_s$ of $G(k)$, the objects $N_i := N \otimes \mathcal{L}_{\chi_i}$ have*

$$G_{geom,\oplus_{i=1}^s N_i} = G_{arith,\oplus_{i=1}^s N_i} = \prod_{i=1}^s Sp(n+1).$$

For the next example, we continue to work on $\mathbb{G}_m/k$ with k of odd characteristic. Recall that a polynomial $f(x) = \sum_{i=0}^{2g} a_i x^i \in k[x]$ of even degree $2g$ is said to be palindromic if $a_{g+i} = a_{g-i}$ for $0 \leq i \leq g$.

Theorem 18.7. *Let $2g \geq 2$ be an even integer, and $f(x) \in k[x]$ a palindromic polynomial of degree $2g$. Suppose that f has $2g$ distinct roots in $\overline{k}$, and that f is not a polynomial in x^d for any prime to p integer $d \geq 2$. Denote by $j : \mathbb{G}_m[1/f] \subset \mathbb{G}_m$ the inclusion. Then both of the objects*

$$N := j_\star \mathcal{L}_{\chi_2(f(x))}(1/2)[1]$$

and

$$M := j_\star \mathcal{L}_{\chi2(xf(x))}(1/2)[1]$$

in $\mathcal{P}_{arith}$ are geometrically irreducible, pure of weight zero, symplecti-cally self-dual of "dimension" $2g$, and have $G_{geom} = G_{arith} = Sp(2g)$.

Proof. Since f is palindromic of degree $2g$ its constant term is nonzero. So all its $2g$ distinct zeroes lie in $\mathbb{G}_m(\overline{k})$. So both $j_\star \mathcal{L}_{\chi2(f(x))}$ and $j_\star \mathcal{L}_{\chi2(xf(x))}$ are irreducible middle extension sheaves on $\mathbb{G}_m$, pure of weight zero and not geometrically isomorphic to any Kummer sheaf $\mathcal{L}_\chi$. So both M and N are geometrically irreducible objects of $\mathcal{P}_{arith}$, pure of weight zero and of "dimension" $2g$. We have

$$M = N \otimes \mathcal{L}_{\chi2}.$$

So $K \mapsto K \otimes \mathcal{L}_{\chi2}$ is a Tannakian isomorphism from $<N>_{arith}$ to $<M>_{arith}$. So N and M have the same groups G_{geom} as each other, and the same groups G_{arith} as each other.

Let us first show that N is geometrically Lie-irreducible, i.e., not geometrically isomorphic to any nontrivial multiplicative translate of itself. This amounts to the statement that $j_\star \mathcal{L}_{\chi2(f(x))} = j_! \mathcal{L}_{\chi2(f(x))}$ is not geometrically isomorphic to $j_! \mathcal{L}_{\chi2(f(ax))}$ for any $a \neq 1$ in $\overline{k}^\times$. We argue by contradiction. Now $j_! \mathcal{L}_{\chi2(f(x))}$ has singularities precisely at the $2g$ zeroes of $f(x)$, while $j_! \mathcal{L}_{\chi2(f(ax))}$ has singularities precisely at the $2g$ zeroes of $f(ax)$. So if the two are geometrically isomorphic for some $a \neq 1$ in $\overline{k}^\times$, then $f(x)$ and $f(ax)$ have the same zeroes. Therefore for some constant $b \in \overline{k}^\times$, we have $f(ax) = bf(x)$. Comparing the nonzero constant terms, we see that $b = 1$, i.e., $f(x) = f(ax)$. But for $d > 1$ the multiplicative order of $a \in \overline{k}^\times$, the equality $f(x) = f(ax)$ implies that $f(x)$ is a polynomial in x^d.

To see that N, or equivalently M, is symplectically self-dual, we remark that for any palindromic $f(x)$ of degree $2g$ in $k[x]$, $f(x)/x^g$ is a palindromic Laurent polynomial of bidegree $(-g, g)$ in $k[x, 1/x]$, so there is a unique polynomial $h_g(x) \in k[x]$ of degree g such that $f(x)/x^g = h_g(x+1/x)$. Because f has $2g$ distinct zeroes, h_g must have g distinct zeroes, none of which is ± 2. If g is even, then

$$j_\star \mathcal{L}_{\chi2(f(x))} = j_\star \mathcal{L}_{\chi2(f(x)/x^g)} = j_\star \mathcal{L}_{\chi2(h_g(x+1/x))},$$

and we apply Theorem 11.1, with the orthogonally self-dual $\mathcal{F} = \mathcal{L}_{\chi2(h_g(x))}$, to conclude that N is symplectically self-dual . [This sheaf $\mathcal{F} = \mathcal{L}_{\chi2(h_g(x))}$ has g distinct singularities in $\mathbb{A}^1$, so at least $g-1$ in $\mathbb{G}_m$, so is not geometrically isomorphic to any Kummer sheaf $\mathcal{L}_\chi$.] If g is odd, then

$$j_\star \mathcal{L}_{\chi 2(xf(x))} = j_\star \mathcal{L}_{\chi 2(f(x)/x^g)} = j_\star \mathcal{L}_{\chi 2(h_g(x+1/x))},$$

and we conclude now by Theorem 11.1 that M is symplectically self-dual. [If $g = 1$, then the unique zero of h_g lies in $\mathbb{G}_m(k)$, for otherwise $f(x)$ would be a constant multiple of $1 + x^2$. So again in this case $\mathcal{F} = \mathcal{L}_{\chi 2(h_g(x))}$ is not geometrically isomorphic to any Kummer sheaf $\mathcal{L}_\chi$.]

The local monodromies of N at 0 and at ∞ are both $Unip(1)$ (because f is invertible at 0 and has even degree), so exactly as in the proof of the previous theorem, $Frob_{k,1}$ gives rise to a Frobenius torus $Diag(x, 1/x, 1, ..., 1)$ in $G^0_{geom,N}$, which is an irreducible subgroup of $Sp(2g)$. So again by the theorem of Kostant and Zarhin [**Ka-ESDE,** 1.2], we conclude that $G^0_{geom,N} = Sp(2g)$, which then forces $G^0_{geom,N} = G_{geom,N} = G_{arith,N} = Sp(2g)$.

$\square$

Again here we can apply Theorem 18.2.

Theorem 18.8. *For $N := j_\star \mathcal{L}_{\chi 2(f(x))}(1/2)[1]$ satisfying all the hypotheses of Theorem 18.7, and for any $s \geq 2$ distinct characters $\chi_1, ..., \chi_s$ of $G(k)$, the objects $N_i := N \otimes \mathcal{L}_{\chi i}$ have*

$$G_{geom, \oplus_{i=1}^s N_i} = G_{arith, \oplus_{i=1}^s N_i} = \prod_{i=1}^s Sp(n).$$

For the next example, we will apply Theorem 10.1. We continue to work on $\mathbb{G}_m/k$.

Theorem 18.9. *Choose a nontrivial character ρ of $k^\times$, of order $r > 2$. Let $f(x) \in k[x]$ be a polynomial whose degree rn is a multiple of r, such that*

(1) *$f(x)$ has rn distinct zeroes in $\overline{k}^\times$.*
(2) *$f(x)$ and $f(1/x)$ have no common zeroes.*
(3) *$f(x)$ is not a polynomial in x^d for any prime to p integer $d > 1$.*

For $j : \mathbb{G}_m[1/f(x)f(1/x)] \subset \mathbb{G}_m$, the object $N := j_\star \mathcal{L}_{\rho(f(x)/f(1/x))}(1/2)[1]$ in $\mathcal{P}_{arith}$ is geometrically irreducible, pure of weight zero, symplectically self-dual of "dimension" $2rn$, and has $G_{geom} = G_{arith} = Sp(2rn)$.

Proof. The object N in $\mathcal{P}_{arith}$ is geometrically irreducible, pure of weight zero, and of "dimension" $2rn$. We next show that it is geometrically Lie-irreducible, i.e., not geometrically isomorphic to any nontrivial multiplicative translate of itself. We argue by contradiction. The object N has singularities precisely at the zeroes of $f(x)$, where its local monodromy is $\mathcal{L}_\rho$, and at the zeroes of $f(1/x)$, where its local monodromy is $\mathcal{L}_{\bar\rho}$. Because ρ has order $r \geq 3$, we recover the zeroes of $f(x)$ as being the singularities where the local monodromy is $\mathcal{L}_\rho$. So if multiplicative translation by some $a \neq 1$ in $\bar k^\times$ preserves the geometric isomorphism class of N, then $f(x)$ and $f(ax)$ have the same zeroes, so are proportional. Because f has a nonzero constant term, we must have $f(ax) = f(x)$. Thus if a has multiplicative order $d \geq 2$, then $f(x)$ is a polynomial in x^d, contradiction.

To apply Theorem 10.1, take for $\mathcal{F}$ there the middle extension of $\mathcal{L}_{\rho(f(x))}$. This shows that N is symplectically self-dual. By the geometric Lie-irreducibility of N, $G^0_{geom,N} \subset Sp(2rn)$ is a connected irreducible subgroup, so an irreducible connected semisimple subgroup. Because f has degree multiple of r, the order of ρ, the local monodromy of $\mathcal{L}_{\rho(f(x)/f(1/x))}$ at both 0 and ∞ is $Unip(1)$, so exactly as in the proof of Theorem 18.1 we get a Frobenius torus $Diag(x, 1/x, 1, ..., 1)$ in $G^0_{geom,N}$. Exactly as in the proof of Theorem 18.2, the theorem of Kostant and Zarhin [**Ka-ESDE**, 1.2] shows that $G^0_{geom,N} = Sp(2rn)$, which then forces $G^0_{geom,N} = G_{geom,N} = G_{arith,N} = Sp(2rn)$. □

Remark 18.10. What becomes of the theorem above if we try to take for ρ the quadratic character χ_2? Thus $f(x)$ has even degree $2d$, and we are looking at $\mathcal{L}_{\chi_2(f(x)/f(1/x))} = \mathcal{L}_{\chi_2(f(x)f(1/x))} = \mathcal{L}_{\chi_2(f(x)x^{2d}f(1/x))} = \mathcal{L}_{\chi_2(f(x)f^{pal}(x))}$. This is a situation to which Theorem 18.2 would apply, to the palindromic polynomial $f(x)f^{pal}(x)$ of degree $4d$, provided that $f(x)f^{pal}(x)$ is not a polynomial in x^e for any integer $e > 1$ prime to p. This condition is strictly stronger than the hypothesis that $f(x)$ not be a polynomial in x^e for any integer $e > 1$ prime to p. For a simple example, take an f for which $f^{pal}(x) = \pm f(-x)$, e.g., $f(x) = x^2 + bx - 1$ with $b \neq 0, b^2 + 4 \neq 0$.

We can apply Theorem 18.2.

Theorem 18.11. *For* $N := j_\star \mathcal{L}_{\rho(f(x)/f(1/x))}(1/2)[1]$ *satisfying all the hypotheses of Theorem 18.9, and for any* $s \geq 2$ *distinct characters* $\chi_1, ..., \chi_s$ *of* $G(k)$*, the objects* $N_i := N \otimes \mathcal{L}_{\chi_i}$ *have*

$$G_{geom,\oplus_{i=1}^s N_i} = G_{arith,\oplus_{i=1}^s N_i} = \prod_{i=1}^s Sp(2rn).$$

We next give an example based on hypergeometric sheaves. We work on $\mathbb{G}_m/k$, with k of odd characteristic. We fix an odd integer $2k+1 \geq 3$, and we consider the hypergeometric sheaf of type $(1, 2k+1)$

$$\mathcal{H} := \mathcal{H}(\psi; \chi_2; \mathbb{1}, ..., \mathbb{1})((2k+1)/2).$$

This $\mathcal{H}$ is lisse on $\mathbb{G}_m$, pure of weight zero and orthogonally self-dual. Its local monodromy at ∞ is $Unip(2k+1)$. Its local monodromy at 0 is the direct sum

$$\mathcal{H}(0) \cong \mathcal{L}_{\chi_2} \oplus Wild_{2k},$$

with $Wild_{2k}$ totally wild of rank $2k$ and Swan conductor 1. This object is Lie-irreducible, and in fact one knows that both its G_{geom} and its G_{arith} are the full orthogonal group $O(2k+1)$. We next form the pullback $\pi^{\star}\mathcal{H}$ of $\mathcal{H}$ by the finite map of degree 2

$$\pi : \mathbb{G}_m \setminus \{1\} \to \mathbb{G}_m, \ x \mapsto x + 1/x - 2.$$

This covering (extended to a finite flat cover of $\mathbb{P}^1$) is finite étale over ∞, with 0 and ∞ the two points in the source lying over it. The covering is doubly ramified over 0, with 1 the unique point in the source lying over it. It is also doubly ramified over -4, with -1 the unique point lying over. For $j_1 : \mathbb{G}_m \setminus \{1\} \subset \mathbb{G}_m$ the inclusion, we form the object $N \in \mathcal{P}_{arith}$ given by

$$N := j_{1\star}\pi^{\star}\mathcal{H}(1/2)[1].$$

This object is pure of weight zero and geometrically irreducible. It is symplectically self-dual, by Theorem 11.1. [View $\pi^{\star}\mathcal{H}$ as the pullback by $x \mapsto x + 1/x$ of the sheaf $[x \mapsto x - 2]^{\star}\mathcal{H}$, and take the sheaf $\mathcal{F}$ of Theorem 11.1 to be the sheaf $[x \mapsto x - 2]^{\star}\mathcal{H}$.]

Theorem 18.12. *The object $N \in \mathcal{P}_{arith}$ is pure of weight zero, and has $G^0_{geom,N} = G_{geom,N} = G_{arith,N} = Sp(2k+2)$.*

Proof. Because 1 is the unique singularity of N in $\mathbb{G}_m(\overline{k})$, N is not geometrically isomorphic to any nontrivial multiplicative translate of itself. Its local monodromy at both 0 and ∞ is $Unip(2k+1)$. Because 1 maps doubly to 0, its local monodromy at 1 is the direct sum of $Unip(1)$ and of a totally wild part of rank $2k$ and Swan conductor 2. So its "dimension" is $drop_1 + Swan_1 = 2k+2$. The result now follows from Theorem 18.1. $\square$

Applying Theorem 18.2 to N, we get the following.

Theorem 18.13. *For any $s \geq 2$ distinct characters $\chi_1, ..., \chi_s$ of $G(k)$, the objects $N_i := N \otimes \mathcal{L}_{\chi_i}$ have*

$$G_{geom,\oplus_{i=1}^s N_i} = G_{arith,\oplus_{i=1}^s N_i} = \prod_{i=1}^s Sp(2k+2).$$

Remark 18.14. Here is a nagging open problem. Suppose we are given $f(x) = \sum_{i=0}^d A_i x^i \in k[x]$ a polynomial of degree d prime to p, which is Artin-Schreier reduced, i.e., A_i vanishes if $p|i$, and such that $gcd\{i|A_i \neq 0\} = 1$. According to Theorem 10.1, the object $N := \mathcal{L}_{\psi(f(x)-f(1/x))}(1/2)[1]$ is symplectically self-dual. The $gcd = 1$ hypothesis, together with Artin-Schreier reducedness, insures that N is not geometrically isomorphic to any nontrivial multiplicative translate of itself, so N is Lie-irreducible. Because N is totally wildly ramified at both 0 and ∞, it has no bad characters χ. We believe that $G_{geom,N} = Sp(2d)$, but in the absence of any bad characters χ, we are unable to prove it (except in the case $d = 1$, the Evans example, where there is a dearth of Lie-irreducible subgroups of $SL(2)$, cf. Theorem 14.2).

This can be viewed as a special case of the following problem. Take for $\mathcal{F}$ a geometrically irreducible lisse sheaf $\mathcal{F}$ on $\mathbb{G}_m/k$ of rank $n \geq 1$ which is pure of weight zero, whose $I(0)$-representation is $Unip(n)$, and whose $I(\infty)$-representation is totally wild, say of Swan conductor $k \geq 1$, and both irreducible and not geometrically isomorphic to any nontrivial multiplicative translate of itself. Denote by $\overline{\mathcal{F}}$ its "complex conjugate," i.e., its linear dual, and form the lisse sheaf

$$\mathcal{G} := \mathcal{F} \otimes ([x \mapsto 1/x]^\star \overline{\mathcal{F}}).$$

Then the local monodromy of $\mathcal{G}$ at both 0 and ∞ is of the form $Unip(n) \otimes Wild_{n,k}$, where $Wild_{n,k}$ is irreducible of rank n and Swan conductor k, and not isomorphic to any nontrivial multiplicative translate of itself. It follows [**Ka-RLS**, 3.1.7] that each of these local monodromies is indecomposable. From this indecomposability, it follows that $\mathcal{G}$ is indecomposable. But $\mathcal{F}$ is itself geometrically irreducible and hence geometrically semisimple, so $\mathcal{G}$, as the tensor product of two geometrically semisimple lisse sheaves, is itself geometrically semisimple. [Alternatively, $\mathcal{G}$ is geometrically semisimple because it is pure of weight zero.] Being indecomposable as well, $\mathcal{G}$ is geometrically irreducible. Looking at its $I(\infty)$-representation, we see that $\mathcal{G}$ is not geometrically isomorphic to any nontrivial multiplicative translate of itself. Thanks to Theorem 10.1, we know that the object $N := \mathcal{G}(1/2)[1]$ is symplectically self-dual. Its "dimension" is $2nk$. Is it always the case that $G_{geom,N} = Sp(2nk)$, or does one need to impose additional conditions on $\mathcal{F}$?

Let us consider the special case when $\mathcal{F}$ is a Kloosterman sheaf $Kl_n := Kl_n(\psi; \mathbb{1}, ..., \mathbb{1})((n-1)/2)$. One knows [**Ka-GKM**, 4.1.1 (3)] that the $I(\infty)$-representation of Kl_n is totally wild, of Swan conductor one, and (consequently, cf. [**Ka-GKM**, 4.1.6 (3)]) is not geometrically isomorphic to any nontrivial multiplicative translate of itself. Its $I(0)$-representation is $Unip(n)$. Is it true that for the associated N, we have

$G_{geom,N} = Sp(2n)$? This seems to be an open question, except for the case $n = 1$, where we once again find the Evans example.

We end this chapter with an example inspired by that of Rudnick, cf. Theorem 14.5.

Theorem 18.15. *Let k be a finite field of odd characteristic, $f(x) \in k[x]$ an odd polynomial (i.e., $f(-x) = -f(x)$) of prime-to-p degree $2n - 1$. Denote by $j_1 : \mathbb{G}_m \setminus \{1\} \to \mathbb{G}_m$ the inclusion. Then the object $N := j_{1\star}\mathcal{L}_{\psi(f((x+1)/(x-1)))}(1/2)[1] \in \mathcal{P}_{arith}$ is pure of weight zero and has $G_{geom,N} = G_{arith,N} = Sp(2n)$.*

Proof. The lisse sheaf $\mathcal{L}_{\psi(f((x+1)/(x-1)))}$ on $\mathbb{G}_m \setminus \{1\}$ is wildly ramified at the point 1, with Swan conductor $2n - 1$, so is not geometrically isomorphic to any $\mathcal{L}_\chi$. Thus N is geometrically irreducible, being a middle extension of generic rank one. Because $g(x) := f((x+1)/(x-1))$ satisfies $g(1/x) = -g(x)$, we can write $g(x) = (1/2)g(x) - (1/2)g(1/x)$, and then apply Theorem 10.1 to see that N is symplectically self-dual. As the unique singularity of N is at the point 1, N is not geometrically isomorphic to any nontrivial multiplicative translate of itself. As N is lisse at both 0 and ∞, and of generic rank one, the result now follows from Theorem 18.1. $\square$

Applying Theorem 18.2, we get the following generalization of this last result.

Theorem 18.16. *For the object N of the previous theorem, and any $r \geq 1$ distinct multiplicative characters $\chi_1, ..., \chi_r$ of $k^\times$, define $N_i := N \otimes \mathcal{L}_{\chi_i}$. Then the object $\oplus_i N_i$ has $G_{geom,\oplus_i N_i} = G_{arith,\oplus_i N_i} = \prod_i Sp(2n)$.*

CHAPTER 19

Orthogonal Examples, Especially $SO(n)$ Examples

The orthogonal case is more difficult than the symplectic one because of the need to distinguish between $SO(n)$ and $O(n)$, which we do not in general know how to do. We work on either the split or the nonsplit form. We begin with a lisse sheaf $\mathcal{F}$ on a dense open set $j : U \subset G$ which is geometrically irreducible, pure of weight zero, and not geometrically isomorphic to (the restriction to U of) any Kummer sheaf $\mathcal{L}_\chi$. We denote by $\mathcal{G} := j_\star\mathcal{F}$ its middle extension to G. Then the object $N := \mathcal{G}(1/2)[1] \in \mathcal{P}_{arith}$ is pure of weight zero and geometrically irreducible. The following result is the orthogonal version of Theorem 18.1.

Theorem 19.1. *Suppose that N is not geometrically isomorphic to any nontrivial multiplicative translate of itself, and that N is orthogonally self-dual. Suppose further that for either of the two possible geometric isomorphisms $G/\overline{k} \cong \mathbb{G}_m/\overline{k}$, both $\mathcal{F}(0)^{unip}$ and $\mathcal{F}(\infty)^{unip}$ are single Jordan blocks $Unip(e)$ of the same size $e \geq 1$. For $n := \dim(\omega(N))$ we have*

$$SO(n) \subset G_{geom,N} \subset G_{arith,N} \subset O(n).$$

Proof. The proof is nearly identical to that of Theorem 18.1. We have a priori inclusions

$$G_{geom,N} \subset G_{arith,N} \subset O(n),$$

so it suffices to prove that $G^0_{geom,N} = SO(n)$. We may extend scalars if necessary from k to its quadratic extension k_2, and reduce to the case where G is $\mathbb{G}_m$. The hypothesis that N is not geometrically isomorphic to any nontrivial multiplicative translate of itself insures that N is geometrically Lie-irreducible, i.e., that G^0_{geom} is an irreducible connected subgroup of $SO(n)$. Thus G^0_{geom} is semisimple, cf. the proof of Theorem 18.1.

The local monodromy of N at both 0 and ∞ is $Unip(e)$. Therefore the semisimplification of $Frob_{k,1}$ gives us a Frobenius torus $Diag(x, 1/x, 1, ..., 1)$ in $G_{arith,N}$. This torus normalizes the connected semisimple group $G^0_{geom,N}$. Exactly as in the symplectic case, it follows that $G^0_{geom,N}$ contains the torus $Diag(x, 1/x, 1, ..., 1)$. By the result of Kostant and

Zarhin [**Ka-ESDE**, 1.2], the only irreducible connected semisimple subgroups of $SL(n)$ which contain $Diag(x, 1/x, 1, ..., 1)$ are $SL(n)$, $SO(n)$, and, when n is even, $Sp(n)$. Since we have an a priori inclusion $G^0_{geom,N} \subset SO(n)$, we must have $G^0_{geom,N} = SO(n)$.   □

Here is the orthogonal analogue of Theorem 18.2. Its proof, via Goursat-Kolchin-Ribet [**Ka-ESDE**, 1.8.2] and Theorems 13.2, 13.3 and 13.4, is entirely analogous.

Theorem 19.2. *Suppose N satisfies all the hypotheses of the theorem above. Suppose further that*

(a) *Either $n \geq 3$ is odd, or $n \geq 6$, $n \neq 8$, is even.*

(b) *At either 0 or at ∞ or at both, the entire tame part of the local monodromy is $Unip(e)$, i.e., local monodromy there is the direct sum of $Unip(e)$ and of something totally wild.*

Given $s \geq 2$ distinct characters $\chi_1, ..., \chi_s$ of $G(k)$, form $N_i := N \otimes \mathcal{L}_{\chi_i}$. Denote by $(\prod_{i=1}^s O(n))_{=det's}$ the subgroup of $\prod_{i=1}^s O(n)$ consisting of those elements whose determinants are either all 1 or all -1. We have the following conclusions.

(1) *If $G_{geom,N} = G_{arith,N} = SO(n)$, then*

$$G_{geom, \oplus_{i=1}^s N_i} = G_{arith, \oplus_{i=1}^s N_i} = \prod_{i=1}^s SO(n).$$

(2) *If $G_{geom,N} = SO(n)$ and $G_{arith,N} = O(n)$, then*

$$G_{geom, \oplus_{i=1}^s N_i} = \prod_{i=1}^s SO(n)$$

and

$$G_{arith, \oplus_{i=1}^s N_i} = (\prod_{i=1}^s O(n))_{=det's}.$$

(3) *If $G_{geom,N} = G_{arith,N} = O(n)$, and if $\chi_i(-1) = 1$ for all i, then*

$$G_{geom, \oplus_{i=1}^s N_i} = G_{geom, \oplus_{i=1}^s N_i} = (\prod_{i=1}^s O(n))_{=det's}.$$

Proof. As already noted, $M \mapsto M \otimes \mathcal{L}_{\chi_i}$ is a Tannakian isomorphism of $<N>_{arith}$ with $<N \otimes \mathcal{L}_{\chi_i}>_{arith}$. In particular, the determinants in the Tannakian sense are related by " det "$(N \otimes \mathcal{L}_{\chi_i}) = $ " det "$(N) \otimes \mathcal{L}_{\chi_i}$. But " det "$(N)$ is either δ_1 (case (1)) or $(-1)^{deg} \otimes \delta_1$ (case (2)) or δ_{-1} or $(-1)^{deg} \otimes \delta_{-1}$ (case (3)), each of which is unchanged when we tensor it with any $\mathcal{L}_{\chi_i}$, so long as, in case (3), $\chi_i(-1) = 1$. Case (1) is immediate from the previous result. In case (2), the previous result gives

$G_{geom, \oplus_{i=1}^s N_i} = \prod_{i=1}^s SO(n)$. We get the asserted value for $G_{arith, \oplus_{i=1}^s N_i}$ by observing that it must strictly contain $\prod_{i=1}^s SO(n)$, but lies in $(\prod_{i=1}^s O(n))_{=det's}$. In case (3), we get the asserted value for $G_{geom, \oplus_{i=1}^s N_i}$ by observing that by the previous result it contains $\prod_{i=1}^s SO(n)$, then that it must strictly contain $\prod_{i=1}^s SO(n)$, and finally that it lies in $(\prod_{i=1}^s O(n))_{=det's}$. Then $G_{arith, \oplus_{i=1}^s N_i}$ contains $(\prod_{i=1}^s O(n))_{=det's}$, but also is contained in it.   □

We now turn to the construction of examples in which $G_{geom, N} = SO(n)$. Let us first explain the method we will use to show that $G_{geom,N}$ is $SO(n)$ rather than $O(n)$. Given a finite subgroup $\Gamma \subset G(\overline{k})$, we say that an object $N \in \mathcal{P}_{geom}$ is adapted to Γ if its restriction to the complement of Γ is lisse, i.e., if $N|(G \setminus \Gamma)$ is a lisse sheaf placed in degree -1. In general it is not true that the middle convolution of two objects adapted to Γ is again adapted to Γ. Here is a simple example. Whatever the choice of Γ, any $N \in \mathcal{P}_{geom}$ which is lisse on G is adapted to Γ. For example, take $\Gamma = \{1\}$, and the Artin-Schreier objects $\mathcal{L}_{\psi(x)}(1/2)[1]$ and $\mathcal{L}_{\psi(-a/x)}(1/2)[1]$. Their middle convolution is δ_a, which is not adapted to this Γ unless $a = 1$. We do, however, have the following lemma.

Lemma 19.3. *Suppose $\Gamma \subset G(\overline{k})$ is a finite subgroup, and $N \in \mathcal{P}_{geom}$ is adapted to Γ and geometrically semisimple. Suppose further that the local monodromy of N at both 0 and ∞ is tame. Then every object in $<N>_{geom}$ is adapted to Γ.*

Proof. Every object in $<N>_{geom}$ is a direct summand of a multiple middle convolution of N and its dual $[x \mapsto 1/x]^* DN$, both of which are adapted to Γ and tame at both 0 and ∞. Proceeding by induction on the number of multiple convolutions, we reduce to the following lemma.   □

Lemma 19.4. *Suppose $\Gamma \subset G(\overline{k})$ is a finite subgroup, and N and M in $\mathcal{P}_{geom}$ are adapted to Γ and geometrically semisimple. Suppose further that the local monodromy of N at both 0 and ∞ is tame. Then the middle convolution $N \star_{mid} M$ is adapted to Γ.*

Proof. We reduce immediately to the case when by N and M are geometrically irreducible. If either is punctual, it is δ_γ for some element $\gamma \in \Gamma$. Then middle convolution with it is multiplicative translation by γ, which preserves being adapted to Γ. So it suffices to treat the case where $N = \mathcal{F}[1]$ and $M = \mathcal{G}[1]$ are each middle extension sheaves placed in degree -1, both adapted to Γ, and where $\mathcal{F}$ is tame at both 0 and ∞. The conditions of being lisse outside Γ, and of being tame at both 0 and ∞, are each stable by Verdier duality. So it suffices to show that the ! convolution $N \star_! M$ is lisse outside Γ. For if this is so, then the same statement applied to their Verdier duals DN and DM, shows that $DN \star_! DM$ is lisse outside Γ, and so its Verdier dual $D(DN \star_! DM) \cong N \star_\star M$ is also adapted to Γ. As the middle convolution $N \star_{mid} M$ is the image of $N \star_! M$ in $N \star_\star M$, it too is lisse outside of Γ.

In order to show that $N \star_! M$ is lisse outside of Γ, we apply Deligne's semicontinuity theorem [**Lau-SCCS**, 2.1.2]. For $a \in G \setminus \Gamma$, the sheaf

$$\mathcal{H} := \mathcal{F} \otimes [x \mapsto a/x]^\star \mathcal{G}$$

is lisse outside the $2\#\Gamma$ points of $\Gamma \cup a\Gamma^{-1}$. By Deligne's theorem, it suffices to show that in the formula for its Euler characteristic,

$$-\chi(\mathcal{H}) = Swan_0(\mathcal{H}) + Swan_\infty(\mathcal{H}) + \sum_{b \in \Gamma \cup a\Gamma^{-1}} (drop_b(\mathcal{H}) + Swan_b(\mathcal{H})),$$

each term is independent of the choice of a, so long as a is not in Γ. Because $\mathcal{F}$ is tame at 0, we have $Swan_0(\mathcal{H}) = rank(\mathcal{F})Swan_\infty(\mathcal{G})$. Because $\mathcal{F}$ is tame at ∞, we have $Swan_\infty(\mathcal{H}) = rank(\mathcal{F})Swan_0(\mathcal{G})$. At a point $\gamma \in \Gamma$, we have $drop_\gamma(\mathcal{H}) = drop_\gamma(\mathcal{F})rank(\mathcal{G})$ and $Swan_\gamma(\mathcal{H}) = Swan_\gamma(\mathcal{F})rank(\mathcal{G})$. At a point $a/\gamma \in a\Gamma^{-1}$, we have $drop_{a/\gamma}(\mathcal{H}) = rank(\mathcal{F})drop_\gamma(\mathcal{G})$ and $Swan_{a/\gamma}(\mathcal{H}) = rank(\mathcal{F})Swan_\gamma(\mathcal{G})$. $\square$

Here is a variant, with the same proof.

Lemma 19.5. *Suppose $S, T \subset G(\overline{k})$ are finite nonempty subsets, and $N \in \mathcal{P}_{geom}$ is lisse outside S and geometrically semisimple. Suppose further that the local monodromy of N at both 0 and ∞ is tame. Suppose $M \in \mathcal{P}_{geom}$ is lisse outside T, and geometrically semisimple. Then their middle convolution $N \star_{mid} M$ is lisse outside $ST := \{st, \ s \in S, \ t \in T\}$.*

Here is yet another variant, with the same proof.

Lemma 19.6. *Suppose $S \subset G(\overline{k})$ is a finite nonempty subset, and $N \in \mathcal{P}_{geom}$ is lisse outside S and geometrically semisimple. Suppose further that the local monodromy of N at both 0 and ∞ is tame. Suppose $M \in \mathcal{P}_{geom}$ is lisse on $\mathbb{G}_m$ (so M is $\mathcal{F}[1]$ for a lisse sheaf $\mathcal{F}$ on $\mathbb{G}_m$). Then their middle convolution $N \star_{mid} M$ is lisse on $\mathbb{G}_m$.*

We can apply these results as follows.

Theorem 19.7. *Let $\Gamma \subset G(\overline{k})$ be a finite subgroup. Suppose that 1 is the only element of order dividing 2 in Γ. This condition is automatic in characteristic 2; in odd characteristic it is the condition that -1 not*

be in Γ. Let $\mathcal{F}$ be a lisse sheaf on $G \setminus \Gamma$ which is geometrically irreducible, pure of weight zero, and not geometrically isomorphic to (the
restriction to $G \setminus \Gamma$ of) any Kummer sheaf $\mathcal{L}_\chi$. We denote by $\mathcal{G} := j_\star \mathcal{F}$
its middle extension to G, and by N the object $N := \mathcal{G}(1/2)[1] \in \mathcal{P}_{arith}$,
which is pure of weight zero and geometrically irreducible. Suppose N is
orthogonally self-dual. Suppose that N is not geometrically isomorphic
to any nontrivial multiplicative translate of itself. Suppose further that
for either of the two possible geometric isomorphisms $G/\overline{k} \cong \mathbb{G}_m/\overline{k}$, $\mathcal{F}$
is tame at both 0 and ∞, and both $\mathcal{F}(0)^{unip}$ and $\mathcal{F}(\infty)^{unip}$ are single
Jordan blocks $Unip(e)$ of the same size $e \geq 1$. For $n := \dim(\omega(N))$ we
have

$$G_{geom,N} = SO(n).$$

Moreover, either $G_{arith,N} = SO(n)$, or "det"(N) in the Tannakian
sense is arithmetically $(-1)^{deg} \otimes \delta_1$, i.e., for any finite extension field
E/k and any character ρ of $G(E)$, $\det(Frob_{E,\rho}) = (-1)^{deg(E/k)}$.

Proof. By Theorem 19.1, we know that $G_{geom,N}$ is either $SO(n)$ or
$O(n)$. So it suffices to show that "det"(N) in the Tannakian sense
is geometrically trivial. It is a one-dimensional object of $<N>_{geom}$
which has order two, so in odd characteristic it is either δ_1 or δ_{-1}. [In
characteristic 2 it can only be δ_1, and we are done.] By Lemma 19.3,
"det"(N) is adapted to Γ. If -1 is not in Γ, "det"(N) cannot be δ_{-1},
so geometrically it must be δ_1. So arithmetically it must be $\alpha^{deg} \otimes \delta_1$,
with $\alpha = \pm 1$. $\qquad\square$

Here is a variant.

Theorem 19.8. *Let $S \subset G(\overline{k})$ be a finite nonempty subset. For each
integer $d \geq 1$, denote by $S^d \subset G(\overline{k})$ the set of all d-fold products of
elements of S. Let $\mathcal{F}$ be a lisse sheaf on $G \setminus S$ which is geometrically irreducible, pure of weight zero, and not geometrically isomorphic to (the
restriction to $G \setminus \Gamma$ of) any Kummer sheaf $\mathcal{L}_\chi$. We denote by $\mathcal{G} := j_\star \mathcal{F}$
its middle extension to G, and by N the object $N := \mathcal{G}(1/2)[1] \in \mathcal{P}_{arith}$,
which is pure of weight zero and geometrically irreducible. Suppose N is
orthogonally self-dual. Suppose that N is not geometrically isomorphic
to any nontrivial multiplicative translate of itself. Suppose further that
for either of the two possible geometric isomorphisms $G/\overline{k} \cong \mathbb{G}_m/\overline{k}$, $\mathcal{F}$
is tame at both 0 and ∞, and both $\mathcal{F}(0)^{unip}$ and $\mathcal{F}(\infty)^{unip}$ are single
Jordan blocks $Unip(e)$ of the same size $e \geq 1$. Suppose further that for
$n := \dim(\omega(N))$ the "dimension" of N, the set S^n does not contain
-1. Then*

$$G_{geom,N} = SO(n).$$

Moreover, either $G_{arith,N} = SO(n)$, or "det"(N) in the Tannakian sense is arithmetically $(-1)^{deg} \otimes \delta_1$, i.e., for any finite extension field E/k and any character ρ of $G(E)$, $\det(Frob_{E,\rho}) = (-1)^{deg(E/k)}$.

Proof. The determinant, in the Tannakian sense, of N is a summand of the n-fold middle convolution of N with itself, so by Lemma 19.5 it is lisse outside S^n. Therefore it cannot be δ_{-1}, and we conclude as in the proof of the previous theorem. $\qquad\square$

We now give some examples. For the first example, we work in odd characteristic. We begin with an irreducible hypergeometric sheaf $\mathcal{H}$ of type $(2m, 2n)$ with $2m < 2n$, of the form

$$\mathcal{H}(\psi; \chi_1, ..., \chi_{2m}; 1, 1, ..., 1)((2m + 2n - 1)/2)$$

which is symplectically self-dual. Its local monodromy at ∞ is $Unip(2n)$, cf. [**Ka-ESDE**, 8.4.11]. Given that all the characters at ∞ are imposed to be trivial, the geometric irreducibility means that no χ_i is trivial, and the symplectic autoduality then means that an even number $2r \geq 0$ of them are the quadratic character, and that the remaining ones, if any, occur in complex conjugate pairs, cf. [**Ka-ESDE**, 8.8.1, 8.8.2]. One knows its geometric monodromy group is $Sp(2n)$, cf. [**Ka-GKM**, 11.6].

Theorem 19.9. *Starting with $\mathcal{H}$ as above, in which $2r$ quadratic characters occur, form the lisse sheaf $\mathcal{G} := [x \mapsto x - 2 + 1/x]^\star \mathcal{H}$ on $\mathbb{G}_m \backslash \{1\}$, then form the object $N := j_{1\star}\mathcal{G}(1/2)[1] \in \mathcal{P}_{arith}$, which is pure of weight zero. Then N is orthogonally self-dual, and we have the following results.*

(1) *If $2r = 0$, then "$\dim$"$(N) = 2n + 2$, and $G_{geom,N} = SO(2n + 2)$.*

(2) *If $2r > 0$, then "$\dim$"$(N) = 2n + 1$, and $G_{geom,N} = SO(2n + 1)$.*

Proof. The pullback $\mathcal{G}$ has local monodromy $Unip(2n)$ at both 0 and ∞. To analyze its local monodromy at 1, observe that the map $x \mapsto x + 1/x - 2$ is doubly ramified over 0, with 1 as the unique point lying over. So the local monodromy of $\mathcal{G}$ at 1 is the direct sum of three pieces: $Unip(2r)$, a tame part of rank $2m - 2r$ with no nonzero inertial invariants, and a totally wild part of rank $2n - 2m$ and Swan conductor 2. So the "dimension" of N is $drop_1 + Swan_1$.

If $2r = 0$, then $drop_1 = 2n$, otherwise $drop_1 = 2n - 1$. Thus the "dimension" is as asserted. By Theorem 11.1, N is orthogonally self-dual. Because 1 is the only singularity of N in $\mathbb{G}_m$, N is not geometrically isomorphic to any nontrivial multiplicative translate of itself. The result now follows from the $\Gamma = \{1\}$ case of Theorem 19.7.   □

In the next example, in odd characteristic $p \neq 5$, we begin with an odd symmetric power of $TwLeg$, say $\mathcal{F} := \mathrm{Sym}^{2d-1}(TwLeg)(d)$. This is a lisse sheaf on $\mathbb{G}_m \setminus \{1\}$ which is pure of weight zero, symplectically self-dual, and has geometric monodromy $SL(2)$, acting in $\mathrm{Sym}^{2d-1}(std_2)$. Its local monodromies at 0 and ∞ are both $Unip(2d)$. Its local monodromy at 1 is $\mathcal{L}_{\chi_2} \otimes Unip(2d)$. We form its pullback $\mathcal{G} := [x \mapsto x + 1/x - 2]^\star\mathcal{F}$, which is lisse on the open set $j : \mathbb{G}_m \setminus \{1, (3\pm\sqrt{5})/2\} \subset \mathbb{G}_m$. Its local monodromy at 1 is $Unip(2d)$. Its local monodromy at each of the points $(3 \pm \sqrt{5})/2$ is $\mathcal{L}_{\chi_2} \otimes Unip(2d)$. We then form

$$N := j_\star\mathcal{G}(1/2)[1],$$

which is pure of weight zero and of "dimension" $drop_1 + drop_{(3+\sqrt{5})/2} + drop_{(3-\sqrt{5})/2} = 2d-1+2d+2d = 6d-1$. By Theorem 11.1, N is orthogonally self-dual. Because 1 is the unique singularity in $\mathbb{G}_m$ at which the local monodromy is unipotent, N is not geometrically isomorphic to any nontrivial multiplicative translate of itself. By Theorem 19.1, we have $SO(6d - 1) \subset G_{geom,N}$.

We wish to apply Theorem 19.8. Here the set S in $\overline{\mathbb{F}_p}^\times$ (remember $p \neq 2, 5$) is the three element set $\{1, (3 + \sqrt{5})/2, (3 - \sqrt{5})/2\}$. Notice that $(3 + \sqrt{5})/2$ is a totally positive unit in the ring of integers of $\mathbb{Q}(\sqrt{5})$, whose inverse is the totally positive unit $(3-\sqrt{5})/2$. This total positivity shows that in $\mathbb{Q}(\sqrt{5})^\times$, the multiplicative subgroup generated by $(3+\sqrt{5})/2$ does not contain -1. It follows that in large characteristic p, the set S^{6d-1} does not contain -1.

To make this precise, let us fix a totally positive unit $u \neq 1$ in the ring of integers $\mathcal{O}_K$ of a real quadratic field K. We define a sequence of strictly positive integers[1] $N(u, n), n \geq 1$, by

$$N(u, n) := (1 + 1)\prod_{i=1}^{n}((u^i + 1)(u^{-i} + 1)) = 2\prod_{i=1}^{n}(\mathrm{Trace}(u^i) + 2).$$

Notice that $N(u, n+1) = (\mathrm{Trace}(u^{n+1}) + 2)N(u, n)$, so the $N(u, n)$ successively divide each other. For our $S = \{1, (3+\sqrt{5})/2, (3-\sqrt{5})/2\}$,

---

[1]The interest of these integers $N(u, n)$ is this. If a prime p does not divide $N(u, n)$, then for any field k of characteristic p, and for any ring homomorphism $\phi : \mathcal{O}_K \to k$, none of the elements $\phi(u^i)$, for $-n \leq i \leq n$, is equal to -1 in k: if p divides $N(u, n)$, then for every such ϕ at least one of the elements $\phi(u^i)$, $-n \leq i \leq n$, is equal to -1 in k.

S^n is the set $\{u^i, -n \leq i \leq n\}$ for $u = (3 + \sqrt{5})/2$. So we get the following theorem.

Theorem 19.10. *Let p be a prime which does not divide the integer $N((3+\sqrt{5})/2, 6d-1)$. Then N as above, formed out of the pullback of $\text{Sym}^{2d-1}(TwLeg)(d)$ by $x \mapsto x + 1/x - 2$, has $G_{geom,N} = SO(6d-1)$.*

Remark 19.11. We do not know if the restriction on p in the above theorem is in fact necessary. For example, starting with $\text{Sym}^1(TwLeg)$, i.e., $d = 1$, we are omitting primes dividing $N((3 + \sqrt{5})/2, 5) = 11025000 = 2^2 3^2 5^5 7^2$. Are the omissions of 3 and 7 needed? We also do not know whether or not we also have $G_{arith,N} = SO(6d-1)$ for the "good" primes p. These same problems persist for the next example as well.

Here is another example, still in odd characteristic, this time based on pulling back a hypergeometric sheaf $\mathcal{H}$ of type $(2n, 2n)$ which is symplectically self-dual. We assume $\mathcal{H}$ is of the form

$$\mathcal{H}(\psi; \chi_1, ..., \chi_{2n}; \mathbb{1}, \mathbb{1}, .., \mathbb{1})((4n - 1)/2),$$

with no χ_i trivial, with an even number $2r$ of the χ_i the quadratic character, and with the remaining ones, if any, occurring in complex conjugate pairs.

Theorem 19.12. *Starting with $\mathcal{H}$ as above, in which $2r$ quadratic characters occur, form the lisse sheaf $\mathcal{G} := [x \mapsto x + 1/x - 2]^* \mathcal{H}$ on $\mathbb{G}_m \setminus \{1, (3 \pm \sqrt{5})/2\}$, then form the object $N := j_\star \mathcal{G}(1/2)[1] \in \mathcal{P}_{arith}$, which is pure of weight zero. Then N is orthogonally self-dual, and we have the following results.*

(1) If $2r = 0$, then "$\dim$"$(N) = 2n + 2$, and $SO(2n + 2) \subset G_{geom,N}$. If $p := char(k)$ does not divide $N((3+\sqrt{5})/2, 2n+2)$, then $G_{geom,N} = SO(2n + 2)$.

(2) If $2r > 0$, then "$\dim$"$(N) = 2n + 1$, and $SO(2n + 1) \subset G_{geom,N}$. If $p := char(k)$ does not divide $N((3+\sqrt{5})/2, 2n+1)$, then $G_{geom,N} = SO(2n + 1)$.

Proof. The pullback $\mathcal{G}$ has local monodromy $Unip(2n)$ at both 0 and ∞. To analyze its local monodromy at 1, observe that the map $x \mapsto x + 1/x - 2$ is doubly ramified over 0, with 1 as the unique point lying over. So the local monodromy of $\mathcal{G}$ at 1 is the direct sum of two (or one, if $2r = 0$) pieces: $Unip(2r)$, and a tame part of rank $2n - 2r$ with no nonzero inertial invariants. At each of the two points $(3 \pm \sqrt{5})/2$, which map to 1, the local monodromy is a unipotent pseudoreflection. So the "dimension" of N is $drop_1 + drop_{(3+\sqrt{5})/2} + drop_{(3-\sqrt{5})/2}$. The first term, $drop_1$, is $2n$ if $2r = 0$, otherwise it is $2n - 1$. At each of the two points $(3 \pm \sqrt{5})/2$, the drop is 1. So the dimension is as asserted. By Theorem 11.1, N is orthogonally self-dual. Because 1 is the unique singularity in $\mathbb{G}_m$ at which the local monodromy is not a unipotent pseudoreflection, N is not geometrically isomorphic to any nontrivial multiplicative translate of itself. So by Theorem 19.1, $G_{geom,N}$ contains the group $SO(2n + 1)$ when $2r > 0$, respectively the group $SO(2n + 2)$ when $2r = 0$. To get the more precise statement, we apply Theorem 19.8. Here the set S is again the three element set $\{1, (3 + \sqrt{5})/2, (3 - \sqrt{5})/2\}$. If the characteristic p does not divide the integer $N((3 + \sqrt{5})/2, 2n + 1)$ when $2r > 0$, respectively the integer $N((3 + \sqrt{5})/2, 2n + 2)$ when $2r > 0$, then S^{2n+1}, respectively S^{2n+2} does not contain -1. □

Remark 19.13. Whenever we have an N for which we know that $G_{geom,N} = SO(n)$ and we know that $G_{arith,N} \subset O(n)$, then either $G_{arith,N} = SO(n)$, or $G_{arith,N} = O(n)$. In the latter case, "det"(N) is geometrically trivial, and of order two, so necessarily $(-1)^{deg}$. [So this latter case cannot arise in characteristic two.] The problem is that in general we don't know which situation we are in. But in both cases, if we extend scalars from the given ground field k to its quadratic extension, we achieve a situation in which $G_{geom} = G_{arith} = SO(n)$.

CHAPTER 20

$GL(n) \times GL(n) \times ... \times GL(n)$ **Examples**

In this chapter, we investigate the following question. Suppose we have a geometrically irreducible middle extension sheaf $\mathcal{G}$ on $\mathbb{G}_m/k$ which is pure of weight zero, such that the object $N := \mathcal{G}(1/2)[1] \in \mathcal{P}_{arith}$ has "dimension" n and has $G_{geom,N} = G_{arith,N} = GL(n)$. Suppose in addition we are given $s \geq 2$ distinct characters χ_i of $k^\times$. We want criteria which insure that for the objects

$$N_i := N \otimes \mathcal{L}_{\chi_i},$$

the direct sum $\oplus_i N_i$ has $G_{geom,\oplus_i N_i} = G_{arith,\oplus_i N_i} = \prod_i GL(n)$. Because we have a priori inclusions $G_{geom,\oplus_i N_i} \subset G_{arith,\oplus_i N_i} \subset \prod_i GL(n)$, it suffices to prove that $G_{geom,\oplus_i N_i} = \prod_i GL(n)$. To show this, it suffices to show both of the following two statements.

(1) The determinants in the Tannakian sense "det"$(N \otimes \mathcal{L}_{\chi_i}) :=$ "det"(N_i) have $G_{geom,\oplus_i"det"(N_i)} = \prod_i GL(1)$.

(2) $(G_{geom,\oplus_i N_i})^{0,der} = \prod_i SL(n)$.

We first deal with the Tannakian determinants. As already noted, $M \mapsto M \otimes \mathcal{L}_\chi$ is a Tannakian isomorphism from $<N>_{geom}$ to $<N \otimes \mathcal{L}_\chi>_{geom}$. In particular, the Tannakian determinants satisfy

$$\text{"det"}(N \otimes \mathcal{L}_\chi) = \text{"det"}(N) \otimes \mathcal{L}_\chi.$$

Now "det"(N) is a nonpunctual (because it is of infinite order) one-dimensional object of $<N>_{geom}$, so it is a multiplicative translate of an irreducible hypergeometric $\mathcal{H}(\psi; \rho_a's; \Lambda_b's)[1]$ of some type (n, m), $n, m \geq 0$, $n + m > 0$, cf. [**Ka-ESDE**, 8.5.3]. The irreducibility is equivalent to the condition that, if both n and m are ≥ 1, no ρ_a is any Λ_b, cf. [**Ka-ESDE**, 8.4.2, 8.4.10.1]. We have the following lemma, which we will apply with its M taken to be "det"(N).

Lemma 20.1. *Suppose $M \in \mathcal{P}_{geom}$ is a multipliplicative translate of an irreducible hypergeometric $\mathcal{H}(\psi; \rho_a's; \Lambda_b's)[1]$ of some type (n, m), $n, m \geq 0$, $n + m > 0$. [Thus if both n and m are ≥ 1, no ρ_a is any Λ_b.] Suppose given $s \geq 2$ distinct characters χ_i of $k^\times$, which satisfy the following three conditions.*

(1) *If both n and m are ≥ 1, then for $i \neq j$, no $\chi_i\rho_a$ is any $\chi_j\Lambda_b$.*

(2) If $n > 0$, then for $i \neq j$, no $\chi_i \rho_a$ is any $\chi_j \rho_{a'}$.

(3) If $m > 0$, then for $i \neq j$, no $\chi_i \Lambda_b$ is any $\chi_j \Lambda_{b'}$.

Then with $M_i := M \otimes \mathcal{L}_{\chi_i}$, we have $G_{geom, \oplus_i M_i} = \prod_i GL(1)$.

Proof. We must show that for any nonzero vector $v = (v_1, ..., v_s) \in \mathbb{Z}^s$, the tensor product in the Tannakian sense $M_1^{\otimes v_1} \otimes M_2^{\otimes v_2} ... \otimes M_s^{\otimes v_s}$ is not geometrically trivial. Omitting terms which don't occur, and renumbering, we must show the following two statements.

(1) If $v_1, v_2, ..., v_r$ are ≥ 1, then the tensor product in the Tannakian sense $M_1^{\otimes v_1} ... \otimes M_r^{\otimes v_r}$ is not geometrically trivial.

(2) If $v_1, v_2, ..., v_r$ are ≥ 1 and $v_{r+1}, ..., v_{r+t}$ are ≤ -1, say $v_i = -w_i$ for $r + 1 \leq i \leq r + t$, then then the tensor product in the Tannakian sense $M_1^{\otimes v_1} \otimes M_2^{\otimes v_2} ... \otimes M_r^{\otimes v_r}$ is not geometrically isomorphic to the tensor product in the Tannakian sense $M_{r+1}^{\otimes w_{r+1}} \otimes ... \otimes M_{r+t}^{\otimes w_{r+t}}$.

To see the truth of these statements, recall from [**Ka-ESDE**, 8.3.3 and 8.4.13.1] that given two irreducible hypergeometrics $\mathcal{H}(\psi; \rho_a's; \Lambda_b's)[1]$ and $\mathcal{H}(\psi; \mu_c's; \nu_d's)[1]$ of types (n, m) and (e, f) respectively, so long as no ρ_a is a ν_d and no Λ_b is a μ_c, then their ! convolution maps isomorphically to their $\star$ convolution. This common convolution is their middle convolution, which is the irreducible hypergeometric of type $(n + e, m + f)$ given up to geometric isomorphism by

$$\mathcal{H}(\psi; \rho_a's; \Lambda_b's)[1] \star_{mid} \mathcal{H}(\psi; \mu_c's; \nu_d's)[1]$$

$$\cong \mathcal{H}(\psi; \rho_a's \cup \mu_c's; \Lambda_b's \cup \nu_d's)[1].$$

Recall also [**Ka-ESDE**, 8.2.5] that for any χ we have

$$\mathcal{H}(\psi; \rho_a's; \Lambda_b's)[1] \otimes \mathcal{L}_\chi \cong \mathcal{H}(\psi; \chi\rho_a's; \chi\Lambda_b's)[1].$$

In case (1), hypothesis (1), if relevant, allows us to compute the tensor product in the Tannakian sense $M_1^{\otimes v_1} ... \otimes M_r^{\otimes v_r}$. It is a multiplicative translate of the irreducible hypergeometric of type $(n \sum_i v_i, m \sum_i v_i)$ whose "upstairs" parameters (the tame part of local monodromy at 0) if any are the $\chi_i \rho_a$ with various repetitions, and whose "downstairs" parameters (the tame part of local monodromy at ∞) if any are the $\chi_j \Lambda_b$ with various repetitions, with i, j in $[1, r]$.

In case (2), $M_1^{\otimes v_1} ... \otimes M_r^{\otimes v_r}$ is a multiplicative translate of the hypergeometric of type $(n \sum_i v_i, m \sum_i v_i)$ whose "upstairs" parameters are the $\chi_i \rho_a$ with various repetitions, and whose "downstairs" parameters are the $\chi_j \Lambda_b$ with various repetitions, with i, j in $[1, r]$. And $M_{r+1}^{\otimes w_{r+1}} \otimes ... \otimes M_{r+t}^{\otimes w_{r+t}}$ is a multiplicative translate of the hypergeometric of type $(n \sum_j w_{r+j}, m \sum_j w_{r+j})$ whose "upstairs" parameters are the $\chi_{r+j}\rho_a$ with various repetitions, and whose "downstairs" parameters are the $\chi_{r+j}\Lambda_b$ with various repetitions, with i, j in $[1, t]$. If $n > 0$ (resp. if $m > 0$), then the tame parts of local monodromy at 0 (resp. at ∞) of both sides are nonzero, but by the disjointness hypotheses (2) (resp. (3)), they have completely disjoint characters at 0 (resp. at ∞). So they are not geometrically isomorphic.   □

Here is a simple but striking case, which gives a more compact packaging of the proof in [**Ka-GKM**, 9.3, 9.5] of a result on equidistribution in $(S^1)^r$ of r-tuples of angles of Gauss sums.

Corollary 20.2. *Fix a nontrivial additive character ψ of k, $\mathcal{L}_\psi$ the corresponding Artin-Schreier sheaf, and put $N := \mathcal{L}_\psi(1/2)[1] \in \mathcal{P}_{arith}$. For any $r \geq 1$ distinct characters χ_i of $k^\times$, put $N_i := N \otimes \mathcal{L}_{\chi_i}$. The object $\oplus_i N_i$ has*

$$G_{geom,\oplus_i N_i} = G_{arith,\oplus_i N_i} = \prod_i GL(1).$$

Proof. $\mathcal{L}_\psi$ is $\mathcal{H}(\psi; \mathbb{1}, \emptyset)$, a hypergeometric of type $(1, 0)$, with the only ρ of the previous lemma the trivial character $\mathbb{1}$.   □

Another simple case is this, which, with $a = 1$, gives an equidistribution result in $(S^1)^r$ for r-tuples of angles of Jacobi sums.

Corollary 20.3. *Fix a nontrivial multiplicative character Λ of $k^\times$, and an element $a \in k^\times$. Put $N := \mathcal{L}_{\Lambda(a-x)}(1/2)[1] \in \mathcal{P}_{arith}$. Choose $r \geq 1$ distinct characters χ_i of $k^\times$, put $N_i := N \otimes \mathcal{L}_{\chi_i}$. Suppose that for all $i \neq j$, $\chi_i \neq \Lambda\chi_j$. Then the object $\oplus_i N_i$ has*

$$G_{geom,\oplus_i N_i} = G_{arith,\oplus_i N_i} = \prod_i GL(1).$$

Proof. Indeed, $\mathcal{L}_{\Lambda(a-x)}$ is a multiplicative translate of the hypergeometric $\mathcal{H}(\mathbb{1}, \Lambda)$ of type $(1, 1)$. Here the only ρ of Lemma 20.1 is $\mathbb{1}$, and the fixed Λ is the only Λ.   □

We now turn to the problem of showing that for a given N with $G_{geom,N} = G_{arith,N} = GL(n)$, we have $(G_{geom,\oplus_i N\otimes\mathcal{L}_{\chi_i}})^{0,der} = \prod_i SL(n)$. Put $N_i := N \otimes \mathcal{L}_{\chi_i}$. By Theorem 13.5 (Goursat-Kolchin-Ribet), it suffices to show that for every one-dimensional object $L \in \mathcal{P}_{geom}$, and for $i \neq j$, there is no geometric isomorphism between N_i and $N_j \star_{mid} L$ nor between N_i and $N_j^\vee \star_{mid} L$.

To deal with an L which is punctual, we must show that for $i \neq j$, there is no geometric isomorphism between N_i and any multiplicative translate of either N_j or of $N_j^\vee$.

To deal with an L which is nonpunctual, we consider the generic rank $gen.rk(M)$ of objects $M \in \mathcal{P}_{geom}$. On a dense open set $U \subset \mathbb{G}_m/\overline{k}$, $M|U$ is $\mathcal{F}[1]$ for a lisse sheaf $\mathcal{F}$ on U. The rank of $\mathcal{F}$ on U is by definition $gen.rk(M)$. Clearly two objects of different generic rank cannot be geometrically isomorphic.

Theorem 20.4. *Suppose $N = \mathcal{G}[1] \in \mathcal{P}_{geom}$, with $\mathcal{G}$ an irreducible middle extension sheaf on $\mathbb{G}_m/\overline{k}$. Suppose that either of the following two conditions is satisfied.*

(1) *"dim"$(N) \geq 3$, $\mathcal{G}$ is tame at 0 and ∞, and its local monodromies at 0 and ∞ both satisfy the following condition: if a character χ occurs, it occurs in a single Jordan block.*

(2) *$\mathcal{G}$ is the restriction to $\mathbb{G}_m$ of a sheaf which is lisse on $\mathbb{A}^1/\overline{k}$, totally wild at ∞, with all of its ∞-slopes > 2.*

Then for any nonpunctual one-dimensional object $L \in \mathcal{P}_{geom}$, we have

$$gen.rk(N \star_{mid} L) > gen.rk(N)$$

and

$$gen.rk(N^\vee \star_{mid} L) > gen.rk(N).$$

Proof. An object and its dual have the same generic rank. The dual of $N^\vee \star_{mid} L$ is $N \star_{mid} L^\vee$, and $L^\vee$ is again a nonpunctual one-dimensional object. So it suffices to prove the first inequality, $gen.rk(N \star_{mid} L) > gen.rk(N)$.

We next explain how to calculate the generic rank of our middle convolution $N \star_{mid} L$. The nonpunctual one-dimensional object L is $\mathcal{H}[1]$, for $\mathcal{H}$ a multiplicative translate of an irreducible hypergeometric sheaf, cf. [**Ka-ESDE**, 8.5.3]. Over a dense open set $U \subset \mathbb{G}_m/\overline{k}$, both $N \star_! L$ and $N \star_\star L$ are lisse sheaves placed in degree -1, and of formation compatible with arbitrary change of base on U. So the middle convolution

$$N \star_{mid} L := Image(N \star_! L \to N \star_\star L)$$

is, on U, itself a lisse sheaf placed in degree -1, and of formation compatible with arbitrary change of base on U.

Denote by S and T respectively the singularities of $\mathcal{G}$ and $\mathcal{H}$ in $\mathbb{G}_m$. Fix a point $a \in U(\overline{k})$ which, if both S and T are nonempty, does not lie in the set ST of products. Form the sheaf $\mathcal{K}$ on $\mathbb{G}_m$ defined by

$$\mathcal{K} := \mathcal{G} \otimes [x \mapsto a/x]^* \mathcal{H}.$$

Notice that $\mathcal{K}$ is itself a middle extension sheaf on $\mathbb{G}_m$ (because a does not lie in ST), and denote by $j : \mathbb{G}_m \subset \mathbb{P}^1$ the inclusion. Then the stalks at a of $N \star_! L$, $N \star_\star L$, and $N \star_{mid} L$ respectively are the cohomology groups

$$H^1(\mathbb{P}^1/\overline{k}, j_!\mathcal{K}), \ H^1(\mathbb{P}^1/\overline{k}, Rj_\star\mathcal{K}), \ H^1(\mathbb{P}^1/\overline{k}, j_\star\mathcal{K})$$

respectively. We have a short exact sequence of sheaves on $\mathbb{P}^1$,

$$0 \to j_!\mathcal{K} \to j_\star\mathcal{K} \to (\mathcal{K}(0)^{I(0)})_{pct. \ at \ 0} \oplus (\mathcal{K}(\infty)^{I(\infty)})_{pct. \ at \ \infty} \to 0.$$

We first observe that in the long exact cohomology sequence, for $i \neq 1$, both $H^i(\mathbb{P}^1/\overline{k}, j_!\mathcal{K})(= H^i_c(\mathbb{G}_m/\overline{k}, \mathcal{K}))$ and $H^i(\mathbb{P}^1/\overline{k}, j_\star\mathcal{K})$ vanish. The H^0_c vanishes because $\mathcal{K}$ has no nonzero punctual sections. The H^2_c and the H^0 vanish because $N = \mathcal{G}[1]$ and hence $N^\vee$ both have "dim"$(N) > 2$, so $N^\vee$ is geometrically isomorphic to no multiplicative translate of $[x \mapsto 1/x]^\star L$, which has "dimension" one. The $H^2(\mathbb{P}^1/\overline{k}, j_\star\mathcal{K})$ vanishes because it is a quotient of $H^2_c(\mathbb{P}^1/\overline{k}, j_!\mathcal{K})$.

So the long exact cohomology sequence gives a short exact sequence

$$0 \to \mathcal{K}(0)^{I(0)} \oplus \mathcal{K}(\infty)^{I(\infty)} \to H^1(\mathbb{P}^1/\overline{k}, j_!\mathcal{K}) \to H^1(\mathbb{P}^1/\overline{k}, j_\star\mathcal{K}) \to 0.$$

Thus we obtain the formula

$$gen.rk(N \star_{mid} L) = gen.rk(N \star_! L) - \dim \mathcal{K}(0)^{I(0)} - \dim \mathcal{K}(\infty)^{I(\infty)}.$$

We will now calculate each of the three terms on the right-hand side. The sheaf $\mathcal{K}$ is lisse outside the disjoint union $S \cup a/T$.

$$gen.rk(N \star_! L) = \dim H^1(\mathbb{P}^1/\overline{k}, j_!\mathcal{K}) = -\chi_c(\mathbb{G}_m/\overline{k}, \mathcal{K})$$

$$= Swan_0(\mathcal{K}) + Swan_\infty(\mathcal{K}) + \sum_{s \in S}(drop_s(\mathcal{K}) + Swan_s(\mathcal{K}))$$

$$+ \sum_{b \in a/T}(drop_b(\mathcal{K}) + Swan_b(\mathcal{K})).$$

For $s \in S$, $\mathcal{H}$ is lisse at a/s, so we have

$$drop_s(\mathcal{K}) + Swan_s(\mathcal{K}) = (drop_s(\mathcal{G}) + Swan_s(\mathcal{G}))gen.rk(\mathcal{H}).$$

For $b = a/t \in a/T$, $\mathcal{G}$ is lisse at b, so we have

$$drop_b(\mathcal{K}) + Swan_b(\mathcal{K}) = gen.rk(\mathcal{G})(drop_t(\mathcal{H}) + Swan_t(\mathcal{H})).$$

We first treat case (1). Thus $\mathcal{G}$ is tame at both 0 and ∞. Here we have

$$Swan_0(\mathcal{K}) = gen.rk(\mathcal{G})Swan_\infty(\mathcal{H}),$$

$$Swan_\infty(\mathcal{K}) = gen.rk(\mathcal{G})Swan_0(\mathcal{H}).$$

In this case, the above plethora of formulas gives the relation

$$gen.rk(N \star_! L) = gen.rk(N)\text{"dim"}(L) + \text{"dim"}(N)gen.rk(L).$$

We next derive, in this case, upper bounds for $\dim \mathcal{K}(0)^{I(0)}$ and for $\dim \mathcal{K}(\infty)^{I(\infty)}$. To do this, write

$$\mathcal{H}(0) = \mathcal{H}(0)^{tame} \oplus \mathcal{H}(0)^{tot.\ wild},$$

$$\mathcal{H}(\infty) = \mathcal{H}(\infty)^{tame} \oplus \mathcal{H}(\infty)^{tot.\ wild}.$$

The isomorphism classes of both $\mathcal{H}(\infty)^{tame}$ and $\mathcal{H}(0)^{tame}$ are invariant under multiplicative translation. Define

$$inv\mathcal{H} := [x \mapsto 1/x]^{\star}\mathcal{H}.$$

Then we have

$$\mathcal{K}(0)^{I(0)} \cong (\mathcal{G}(0) \otimes inv\mathcal{H}(0)^{tame})^{I(0)},$$

$$\mathcal{K}(\infty)^{I(\infty)} \cong (\mathcal{G}(\infty) \otimes inv\mathcal{H}(\infty)^{tame})^{I(\infty)}.$$

Now we make use of the hypothesis on $\mathcal{G}$ that its local monodromies at both 0 and ∞ satisfy the condition that any character that occurs does so in a single Jordan block. One knows [**Ka-ESDE**, 8.4.2 (6-8)] that the tame parts of the local monodromies of $\mathcal{H}$ and hence of $inv\mathcal{H}$ at both 0 and ∞ also satisfy this condition. We claim that

$$\dim \mathcal{K}(0)^{I(0)} \leq Min(gen.rk(\mathcal{G}), rk(inv\mathcal{H}(0)^{tame})),$$

$$\dim \mathcal{K}(\infty)^{I(\infty)} \leq Min(gen.rk(\mathcal{G}), rk(inv\mathcal{H}(\infty)^{tame})).$$

Granting these claims, we can conclude as follows. It suffices to show that

$$gen.rk(N)\text{``}\dim\text{''}(L) + \text{``}\dim\text{''}(N)gen.rk(L) - 2Min(gen.rk(\mathcal{G}), gen.rk(\mathcal{H}))$$
$$> gen.rk(N).$$

Now $\text{``}\dim\text{''}(L) = 1$, so this is equivalent to showing that

$$\text{``}\dim\text{''}(N)gen.rk(L) > 2Min(gen.rk(\mathcal{G}), gen.rk(\mathcal{H})).$$

But $gen.rk(L) := gen.rk(\mathcal{H})$, we may rewrite this as

$$\text{``}\dim\text{''}(N)gen.rk(\mathcal{H}) > 2Min(gen.rk(\mathcal{G}), gen.rk(\mathcal{H})).$$

But we have the trivial inequality $gen.rk(\mathcal{H}) \geq Min(gen.rk(\mathcal{G}), gen.rk(\mathcal{H}))$, so it suffices that

$$\text{``}\dim\text{''}(N)gen.rk(\mathcal{H}) > 2gen.rk(\mathcal{H}),$$

which is obvious from the assumption that $\text{``}\dim\text{''}(N) \geq 3$.

It remains to show that

$$\dim \mathcal{K}(0)^{I(0)} \leq Min(gen.rk(\mathcal{G}), rk(inv\mathcal{H}(0)^{tame})),$$

$$\dim \mathcal{K}(\infty)^{I(\infty)} \leq Min(gen.rk(\mathcal{G}), rk(inv\mathcal{H}(\infty)^{tame})).$$

Recall that

$$\mathcal{K}(0)^{I(0)} \cong (\mathcal{G}(0) \otimes inv\mathcal{H}(0)^{tame})^{I(0)},$$

$$\mathcal{K}(\infty)^{I(\infty)} \cong (\mathcal{G}(\infty) \otimes inv\mathcal{H}(\infty)^{tame})^{I(\infty)}.$$

So we must show that

$$\dim(\mathcal{G}(0) \otimes inv\mathcal{H}(0)^{tame})^{I(0)} \leq Min(gen.rk(\mathcal{G}), rk(inv\mathcal{H}(0)^{tame})),$$

$$\dim(\mathcal{G}(\infty) \otimes inv\mathcal{H}(\infty)^{tame})^{I(\infty)} \leq Min(gen.rk(\mathcal{G}), rk(inv\mathcal{H}(\infty)^{tame})).$$

It suffices to prove the (universal truth of the) first. Thus we have a tame representation V of $I(0)$

$$V := \bigoplus_i \mathcal{L}_{\chi_i} \otimes Unip(n_i)$$

of rank $n = \sum_i n_i$, in which the χ_i are all distinct. And we have a second tame representation W of $I(0)$,

$$W := \bigoplus_j \mathcal{L}_{\overline{\rho_j}} \otimes Unip(m_j)$$

of rank $m = \sum_j m_j$, in which all the ρ_j are distinct. We must show that

$$\dim((V \otimes W)^{I(0)}) \leq Min(n, m).$$

Now

$$V \otimes W = \bigoplus_{i,j} \mathcal{L}_{\chi_i/\rho_j} \otimes (Unip(n_i) \otimes Unip(m_j)).$$

The only terms with nonzero $I(0)$-invariants are those for which $\chi_i = \rho_j$. If there are no such pairs, there is nothing to prove. If there are such pairs, we may renumber so that $\chi_i = \rho_i$ for $i = 1, ..., r$, and $\chi_i \neq \rho_j$ unless $i = j$ and $1 \leq i, j \leq r$. So we may replace V by its subspace $\bigoplus_{i=1}^r \mathcal{L}_{\chi_i} \otimes Unip(n_i)$ and replace W by its subspace $\bigoplus_{i=1}^r \mathcal{L}_{\overline{\chi_i}} \otimes Unip(m_i)$. So we are reduced to showing that

$$\sum_{i=1}^r \dim((Unip(n_i) \otimes Unip(m_i))^{I(0)}) \leq Min(\sum_{i=1}^r n_i, \sum_{i=1}^r m_i).$$

We have the trivial inequality

$$\sum_{i=1}^r Min(n_i, m_i) \leq Min(\sum_{i=1}^r n_i, \sum_{i=1}^r m_i).$$

So it suffices to observe that for two integers $n, m \geq 1$, we have

$$\dim((Unip(n) \otimes Unip(m))^{I(0)}) = Min(n, m).$$

To fix ideas, say $m \leq n$. One knows that

$$Unip(n) \otimes Unip(m) = Sym^{n-1}(Unip(2)) \otimes Sym^{m-1}(Unip(2))$$

$$\cong \bigoplus_{i=0}^{m-1} \mathrm{Sym}^{n-1+m-1-2i}(Unip(2)) = \bigoplus_{i=0}^{m-1} Unip(n+m-1-2i),$$

and each of the m unipotent Jordan block summands has a one-dimension space of $I(0)$-invariants. This concludes the proof in case (1).

We now treat case (2). Thus $\mathcal{G}$ is the restriction to $\mathbb{G}_m$ of a sheaf which is lisse on $\mathbb{A}^1/\overline{k}$, totally wild at ∞, with all of its ∞-slopes > 2. Recall that

$$gen.rk(N \star_! L) = \dim H^1(\mathbb{P}^1/\overline{k}, j_! \mathcal{K}) = -\chi_c(\mathbb{G}_m/\overline{k}, \mathcal{K})$$

$$= Swan_0(\mathcal{K}) + Swan_\infty(\mathcal{K}) + \sum_{s \in S}(drop_s(\mathcal{K}) + Swan_s(\mathcal{K}))$$

$$+ \sum_{b \in S \in a/T}(drop_b(\mathcal{K}) + Swan_b(\mathcal{K})).$$

Here S is empty. For $b = a/t \in a/T$, $\mathcal{G}$ is lisse at b, so we have

$$drop_b(\mathcal{K}) + Swan_b(\mathcal{K}) = gen.rk(\mathcal{G})(drop_t(\mathcal{H}) + Swan_t(\mathcal{H})).$$

Because $\mathcal{G}$ is lisse at 0, we have

$$Swan_0(\mathcal{K}) = gen.rk(\mathcal{G})Swan_\infty(\mathcal{H}).$$

Because $\mathcal{G}$ has all its ∞-slopes > 2, while every ∞-slope of $[x \mapsto a/x]^\star \mathcal{H}$ is ≤ 1, we have

$$Swan_\infty(\mathcal{K}) = gen.rk(\mathcal{H})Swan_\infty(\mathcal{G}).$$

Putting all these together, we get the formula

$$gen.rk(N\star_!L) = gen.rk(\mathcal{G})(\text{``dim''}(L) - Swan_0(\mathcal{H})) + \text{``dim''}(N)gen.rk(L).$$

Meanwhile $\mathcal{K}$ is totally wild at ∞, so $\mathcal{K}(\infty)^{I(\infty)} = 0$. As $\mathcal{G}$ is lisse at 0,

$$\dim(\mathcal{K}^{I(0)}) = gen.rk.(\mathcal{G})\dim(inv\mathcal{H}(0)^{I(0)}).$$

Now $\text{``dim''}(L) = 1$, and both $Swan_0(\mathcal{H})$ and $\dim(inv\mathcal{H}(0)^{I(0)})$ are ≤ 1. So we have the inequality

$$gen.rk(N \star_! L) \geq \text{``dim''}(N)gen.rk(L) - gen.rk.(\mathcal{G}).$$

We will show that

$$\text{``dim''}(N)gen.rk(L) - gen.rk.(\mathcal{G}) > gen.rk.(\mathcal{G}),$$

i.e., that

$$\text{``dim''}(N)gen.rk(L) > 2gen.rk.(\mathcal{G}).$$

As $gen.rk(L) \geq 1$, it suffices to show that

$$\text{``dim''}(N) > 2gen.rk.(\mathcal{G}).$$

But $\text{``dim''}(N) = Swan_\infty(\mathcal{G})$ is the sum of the $gen.rk(\mathcal{G})$ ∞-slopes of $\mathcal{G}$, each of which is > 2, so this last inequality is obvious. $\qquad\square$

With these results in hand, it is a simple matter to apply them to the $GL(n)$ examples we have found in Chapter 17.

Theorem 20.5. *Fix any integer $n \geq 3$, and a lisse rank one sheaf $\mathcal{F}$ on $\mathbb{A}^1/\mathbb{F}_p$ which is pure of weight zero and whose Swan conductor at ∞ is the integer n, such that $\mathcal{F}|\mathbb{G}_m$ is not geometrically isomorphic to any nontrivial multiplicative translate of itself. Thus the object $N :=$ $\mathcal{F}(1/2)[1] \in \mathcal{P}_{arith}$ has "dim"$(N) = n$, and $G_{geom,N} = G_{arith,N} = GL(n)$. For any $r \geq 2$ distinct characters χ_i of $k^\times$, put $N_i := N \otimes \mathcal{L}_{\chi_i}$. Then the object $\oplus_i N_i$ has*

$$G_{geom,\oplus_i N_i} = G_{arith,\oplus_i N_i} = \prod_i GL(n).$$

Proof. We first show that for $i \neq j$, there is no geometric isomorphism between N_i and any multiplicative translate of either N_j or $N_j^\vee$. Any multiplicative translate of $N_j^\vee$ is totally wild at 0, while N_i is tame at 0. And N_i and any multiplicative translate of N_j have nonisomorphic local monodromies at 0, namely $\mathcal{L}_{\chi_i}$ and $\mathcal{L}_{\chi_j}$.

To show that for a nonpunctual $L \in \mathcal{P}_{geom}$ of "dimension" one there is no geometric isomorphism of N_i with either $N_j \star_{mid} L$ or $N_j^\vee \star_{mid} L$, we "pull out" the $\mathcal{L}_{\chi_j}$:

$$N_j \star_{mid} L = (N \otimes \mathcal{L}_{\chi_j}) \star_{mid} L \cong (N \star_{mid} (L \otimes \mathcal{L}_{\overline{\chi_j}})) \otimes \mathcal{L}_{\chi_j}.$$

Now N and each N_j have the same generic rank as each other, so by the previous result we get

$$gen.rk(N_j \star_{mid} L) > gen.rk(N_i).$$

Writing

$$N_j^\vee \star_{mid} L = (N^\vee \otimes \mathcal{L}_{\chi_j}) \star_{mid} L \cong (N^\vee \star_{mid} (L \otimes \mathcal{L}_{\overline{\chi_j}})) \otimes \mathcal{L}_{\chi_j},$$

and putting $L_1 := L \otimes \mathcal{L}_{\overline{\chi_j}}$, we observe that $N^\vee \star_{mid} L_1$ has the same generic rank as its Tannakian dual $N \star_{mid} L_1^\vee$, so we get

$$gen.rk(N_j^\vee \star_{mid} L) > gen.rk(N_i).$$

To show that the Tannakian determinants "det"$(N_i) = (\det N) \otimes \mathcal{L}_{\chi_i}$ are independent, it suffices, by Corollary 20.2, to show that "det"(N) is geometrically isomorphic to a multiplicative translate of the Artin Schreier sheaf $\mathcal{L}_\psi(1/2)[1]$, ψ some fixed nontrivial additive character of k. For this we argue as follows. One knows that the collection of all objects in $\mathcal{P}_{geom}$ which are lisse on $\mathbb{G}_m$, unipotent at 0, and totally wild at ∞ is stable by $!$ convolution, which coincides, on this collection, with middle convolution, cf. [**Ka-GKM**, 5.1 (2)]. So "det"(N), a direct factor of the n-fold convolution of N with itself, is lisse on $\mathbb{G}_m$, totally wild at 0, and unipotent at 0. The only such objects of "dimension" one are (shifts by [1] of) multiplicative translates of Kloosterman sheaves $Kl_n := Kl(\psi; \mathbb{1}, ..., \mathbb{1})(n/2)$ of some rank $n \geq 1$, cf. [**Ka-ESDE**, 8.5.3]. Because "det"$(N) \in \mathcal{P}_{arith}$, the multiplicative translation must be by an $a \in k^\times$. It remains to see that $n = 1$. For this we consider the weight drop of $Frob_{k,\mathbb{1}}$ acting on "det"(N). Acting on N, it has $n - 1$ eigenvalues of absolute value 1, and one eigenvalue of absolute value $1/\sqrt{\#k}$. So acting on "det"(N), its weight drop is one. In general, the weight drop of $Frob_{k,\mathbb{1}}$ acting on $\omega(Kl_n(n/2)[1]) \cong H^1_c(\mathbb{G}_m/\overline{k}, Kl_n(n/2))$ is n, cf. [**Ka-GKM**, 7.3.1]. $\qquad\square$

Theorem 20.6. *Let k have odd characteristic. Fix an odd integer $n \geq 3$, and define $N := \mathrm{Sym}^n(Leg)((n+1)/2)[1]$ in $\mathcal{P}_{arith}$, which we have seen (Theorem 17.2) is pure of weight zero, has "dimension" n, and has*

$$G_{geom,N} = G_{arith,N} = GL(n).$$

Let $\chi_1, ..., \chi_r$ be characters of $k^\times$ whose squares are all distinct, i.e., for $i \neq j$, χ_i/χ_j is neither trivial nor is it the quadratic character. Put $N_i := N \otimes \mathcal{L}_{\chi_i}$. Then the object $\oplus_i N_i$ has

$$G_{geom,\oplus_i N_i} = G_{arith,\oplus_i N_i} = \prod_i GL(n).$$

Proof. The objects N_i and their Tannakian duals $N_i^\vee$ all have the point 1 as their unique singularity in $\mathbb{G}_m$, and all are tame at 0 and ∞. So none is geometrically isomorphic to a nontrivial multiplicative translate of another. The local monodromy of N at 0 is $Unip(n+1)$, and that of $N^\vee$ at 0 is $\mathcal{L}_{\chi_{quad}} \otimes Unip(n+1)$, so for $i \neq j$, N_i is not geometrically isomorphic to either N_j or $N_j^\vee$. We may apply Theorem 20.4 to show that no N_i is isomorphic to any $N_j \star_{mid} L$ or to any $N_j^\vee \star_{mid} L$ for any nonpunctual one-dimensional object L. The generic rank method applies, because all the N_i and $N_i^\vee$ have the same generic rank $n+1$, and are tame at 0 and ∞, with local monodromy at each a single Jordan block of size $n + 1$.

It remains to show how to apply Lemma 20.1 to "det"(N). As explained in the proof of Theorem 17.2, the only bad characters for N are the trivial character $\mathbb{1}$ and the quadratic character χ_2. For $\mathbb{1}$, $n - 1$ of the Frobenius eigenvalues have absolute value 1, and the remaining eigenvalue has absolute value $(1/\sqrt{\#k})^{n+1}$. For χ_2, $n - 1$ of the Frobenius eigenvalues have absolute value 1, and the remaining eigenvalue has absolute value $(\sqrt{\#k})^{n+1}$. So "det"(N) is geometrically isomorphic to a multiplicative translate of $\mathcal{H}[1]$ for $\mathcal{H}$ the hypergeometric of type $(n + 1, n + 1)$ given by $\mathcal{H}(\psi; \mathbb{1}, ..., \mathbb{1}; \chi_2, ..., \chi_2)$. And by Lemma 19.3, there is in fact no multiplicative translate. □

We also have the following generalization.

Theorem 20.7. *Let $n \geq 3$ be an integer, and χ a character of $k^{\times}$ such that χ^2 has order $> n$. Form the hypergeometric sheaf $\mathcal{H} := \mathcal{H}(!, \psi; 1, \chi^2; \chi, \chi)$, and the object $N := \mathrm{Sym}^n(\mathcal{H})((3n + 1)/2)[1]$ in $\mathcal{P}_{arith}$, which (by Theorem 17.3) is pure of weight zero, has "dimension" n, and has*

$$G_{geom,N} = G_{arith,N} = GL(n).$$

Let $\rho_1, ..., \rho_r$ be $r \geq 2$ characters of $k^{\times}$, such that for $i \neq j$, ρ_i/ρ_j is not on either of the following two lists:

$$\{\chi^{2n-2j}\}_{j=0,...,2n}, \quad \{\chi^{n-2j}\}_{j=0,...,n}.$$

Put $N_i := N \otimes \mathcal{L}_{\chi_i}$. Then the object $\oplus_i N_i$ has

$$G_{geom,\oplus_i N_i} = G_{arith,\oplus_i N_i} = \prod_i GL(n).$$

Proof. The objects N_i and their Tannakian duals $N_i^{\vee}$ all have the point 1 as their unique singularity in $\mathbb{G}_m$, and all are tame at 0 and ∞. So none is geometrically isomorphic to a nontrivial multiplicative translate of another. The local monodromy of N at 0 is $\oplus_{j=0}^n \mathcal{L}_{\chi^{2j}}$, and that of $N^{\vee}$ at 0 is $\mathcal{L}_{\chi^n} \otimes Unip(n + 1)$. In view of the hypotheses made on the ratios of the ρ_i, we see from the local monodromies at 0 that for $i \neq j$, N_i is not geometrically isomorphic to either N_j or $N_j^{\vee}$. We use Theorem 20.4 to show that no N_i is isomorphic to any $N_j \star_{mid} L$ or to any $N_j^{\vee} \star_{mid} L$ for any nonpunctual one-dimensional object L.

To apply Lemma 20.1 to "det"(N), recall from the proof of Theorem 17.3 that this Tannakian determinant is geometrically isomorphic to $\mathcal{H}[1]$ for $\mathcal{H}$ the hypergeometric sheaf of type (n, n) given by $\mathcal{H}(\psi; \chi^2, ..., \chi^{2n}; \mathbb{1}, ..., \mathbb{1})$. □

To conclude this chapter, we consider our fourth example of a $GL(n)$ object. Recall that in Theorem 17.5 we took a polynomial $f[x] = \sum_{i=0}^n A_i x^i$ in $k[x]$ of degree $n \geq 2$ with all distinct roots in $\bar{k}$. We supposed that $f(0) \neq 0$, and that $\gcd\{i|A_i \neq 0\} = 1$, and we took a character χ of $k^{\times}$ such that χ^n is nontrivial. We then formed the object $N := \mathcal{L}_{\chi(f)}(1/2)[1]$ in $\mathcal{P}_{arith}$, which we showed was pure of weight zero, had "dimension" n, and had

$$G_{geom,N} = G_{arith,N} = GL(n).$$

In order to apply Theorem 20.4, we need to assume $n \geq 3$ in the following theorem.

Theorem 20.8. *Suppose $n \geq 3$, and consider the object $N := \mathcal{L}_{\chi(f)}(1/2)[1]$ of the previous paragraph. Let $\rho_1, ..., \rho_r$ be $r \geq 2$ characters of $k^\times$, such that for $i \neq j$, χ_i/χ_j is neither the trivial character $\mathbb{1}$ nor the character χ^n. Put $N_i := N \otimes \mathcal{L}_{\chi_i}$. Then the object $\oplus_i N_i$ has*

$$G_{geom, \oplus_i N_i} = G_{arith, \oplus_i N_i} = \prod_i GL(n).$$

Proof. Recall from the proof of Theorem 17.5 that in this case the Tannakian determinant "det"(N) is geometrically isomorphic to $\mathcal{H}[1]$ for $\mathcal{H} = \mathcal{H}(\psi; \mathbb{1}; \chi^n) \cong \mathcal{L}_{\chi^n(1-x)}$. So by Lemma 20.1 the determinants of the N_i are independent.

Because we assume $n \geq 3$, we may apply Theorem 20.4 to show that no N_i is isomorphic to any $N_j \star_{mid} L$ or to any $N_j^\vee \star_{mid} L$ for any nonpunctual one-dimensional object L. It remains to show that for $i \neq j$, N_i is not geometrically isomorphic to any multiplicative translate of either N_j or of $N_j^\vee$. To see this, we compare local monodromies at ∞. For N_i it is $\mathcal{L}_{\chi^n \rho_i}$. For N_j it is $\mathcal{L}_{\chi^n \rho_j}$, and for $N_j^\vee$ it is $\mathcal{L}_{\rho_j}$, all of whose geometric isomorphism classes are invariant under multiplicative translation. By the hypothesis on the characters ρ_k, these local monodromies are in three distinct isomorphism classes. $\square$

CHAPTER 21

$SL(n)$ Examples, for n an Odd Prime

In this chapter, we will construct, for every $n \geq 3$, n-dimensional objects with $G_{geom} \subset G_{arith} \subset SL(n)$, but only when n is prime will we be able to prove that $G_{geom} = G_{arith} = SL(n)$.

Theorem 21.1. *Let k be a finite field of characteristic p, ψ a nontrivial additive character of k. Let $f(x) = \sum_{i=-b}^{a} A_i x^i \in k[x, 1/x]$ be a Laurent polynomial of "bidegree" (a, b), with a, b both ≥ 1 and both prime to p. Assume further that $f(x)$ is Artin-Schreier reduced. Thus $A_a A_{-b} \neq 0$, and $A_i \neq 0$ implies that i is prime to p. We have the following results.*

(1) *The object $N := \mathcal{L}_{\psi(f(x))}(1/2)[1] \in \mathcal{P}_{arith}$ is pure of weight zero, geometrically irreducible, and of "dimension" $a + b$.*

(2) *If $(-1)^a a A_a = (-1)^{b+1} b A_{-b}$, then $G_{geom,N} \subset SL(a+b)$. In general, the Tannakian determinant "$\det$"(N) is geometrically isomorphic to δ_α, for $\alpha = (-1)^{b+1} b A_{-b}/(-1)^a a A_a$. And setting*

$$\beta := \det(Frob_{k,\mathbf{1}} | \omega(N)),$$

we have the arithmetic determinant formula

$$\text{``}\det\text{''}(N) = \beta^{deg} \otimes \delta_\alpha.$$

(3) *If $a \neq b$, then N is **not** Lie-self-dual, i.e., $G^0_{geom,N}$ lies in no orthogonal group $SO(a+b)$, and, if $a + b$ is even, it lies in no symplectic group $Sp(a+b)$.*

(4) *Suppose in addition that $\gcd\{i | A_i \neq 0\} = 1$. Then N is geometrically Lie-irreducible, i.e., $G^0_{geom,N}$ is an irreducible subgroup of $SL(a+b)$ (in the given $a + b$-dimensional representation).*

Proof. The sheaf $\mathcal{L}_{\psi(f(x))}$ is lisse of rank one (and a fortiori geometrically irreducible) and pure of weight zero on $\mathbb{G}_m$, with Euler characteristic $-Swan_0 - Swan_\infty = -a - b$, so (1) is obvious. Let us admit the truth of (2) for the moment. To show that N is not Lie-self-dual when $a \neq b$, it suffices to show that for any integer $d \geq 1$ prime to p, the d'th power direct image $[d]_\star N$ is not geometrically isomorphic to its Tannakian dual. But this object has different Swan conductors at 0 and ∞, namely a and b respectively, while its Tannakian dual

$[x \mapsto 1/x]^\star D([d]_\star N)$ has Swan conductors b and a respectively at 0 and ∞. If in addition $gcd\{i | A_i \neq 0\} = 1$, the geometric Lie-irreducibility of N follows from Corollary 8.3. Indeed, this $gcd = 1$ hypothesis and the fact that f is Artin-Schreier reduced together insure that for any $a \neq 1$ in $\overline{k}^\times$, N is not geometrically isomorphic to $[x \mapsto ax]^\star N$, cf. the proof of Theorem 14.2.

We now turn to the calculation of the Tannakian determinant "det"(N) We will compute, for every finite extension field E/k, and every character χ of $E^\times$,

$$\det(Frob_{E,\chi} | \omega(N)) = \det(Frob_E | H^0(\mathbb{A}^1 \otimes_k \overline{k}, j_{0!}(N \otimes \mathcal{L}_\chi))).$$

Because N is totally wild at both 0 and ∞, this is

$$= \det(Frob_E | H^0_c(\mathbb{G}_m \otimes_k \overline{k}, N \otimes \mathcal{L}_\chi))$$

$$= \det(Frob_E | H^1_c(\mathbb{G}_m \otimes_k \overline{k}, \mathcal{L}_{\psi(f(x))} \otimes \mathcal{L}_\chi)(1/2))$$

$$= \sqrt{\#E}^{-a-b} \det(Frob_E | H^1_c(\mathbb{G}_m \otimes_k \overline{k}, \mathcal{L}_{\psi(f(x))} \otimes \mathcal{L}_\chi)).$$

We follow the method of Hasse-Davenport [**D-H**]. The L-function on $\mathbb{G}_m/E$ with coefficients in $\mathcal{L}_{\psi(f(x))} \otimes \mathcal{L}_\chi$ has the additive expression

$$L(\mathbb{G}_m/E, T, \mathcal{L}_{\psi(f(x))} \otimes \mathcal{L}_\chi) = 1 + \sum_{d \geq 1} T^d \sum_{D \in EffDiv^d(\mathbb{G}_m/E)} \psi(D)\chi(D),$$

where the inner sum is over the space of effective divisors of degree d on $\mathbb{G}_m/E$. Concretely, the effective divisors D of given degree d are the monic polynomials of degree d in $E[x]$ of the form $f_D := x^d - s_1 x^{d-1} + \ldots + (-1)^d s_d$, with s_d invertible. The term $\chi(D)$ for this divisor is $\chi(s_d)$. The term $\psi(D)$ for this divisor is $\psi(\sum_{i=-b}^a A_i N_i)$, for N_i the sum of the i'th powers of the roots of the corresponding polynomial f_D. [N.B. Because f_D has an invertible constant term, all its roots are invertible, so N_i makes sense for negative i as well.] Comparing the additive expression for the L-function with its cohomological expression

$$L(\mathbb{G}_m/E, T, \mathcal{L}_{\psi(f(x))} \otimes \mathcal{L}_\chi) = \det(1 - T Frob_E | H^1_c(\mathbb{G}_m \otimes_k \overline{k}, \mathcal{L}_{\psi(f(x))} \otimes \mathcal{L}_\chi)),$$

which shows the L function to be a polynomial in T of degree $a + b$, and equating coefficients of T^{a+b}, we get the identity

$$\det(-Frob_E | H^1_c(\mathbb{G}_m \otimes_k \overline{k}, \mathcal{L}_{\psi(f(x))} \otimes \mathcal{L}_\chi)) = \sum_{D \in EffDiv^{a+b}} \psi(D)\chi(D)$$

$$= \sum_{s_1, \ldots, s_{a+b} \in E, s_{a+b} \neq 0} \chi(s_{a+b}) \psi\left(\sum_{i=-b}^a A_i N_i\right).$$

With the convention that for $i > 0$, s_{-i} is the i'th elementary symmetric function of the inverses of the roots, we have

$$s_{-i} = s_{a+b-i}/s_{a+b}.$$

And by the relation between Newton symmetric functions and elementary symmetric functions, we have

$$N_a = (-1)^{a+1} a s_a + p_a(s_1, ..., s_{a-1}),$$

where $p_a(s_1, ..., s_{a-1})$ is isobaric of weight a and does not involve s_a. Similarly,

$$N_{-b} = (-1)^{b+1} b s_{-b} + p_b(s_{-1}, ..., s_{-(b-1)}),$$

where $p_b(s_{-1}, ..., s_{-(b-1)})$ is isobaric of weight b and does not involve s_{-b}. The terms N_i with $0 \le i < a$ are polynomials in the variables $s_1, ..., s_i$, and the terms N_{-j} with $0 < j < b$ are polynomials in $s_{-1}, ..., s_{-j}$. Thus we get an expression for

$$\det(-Frob_E | H_c^1(\mathbb{G}_m \otimes_k \overline{k}, \mathcal{L}_{\psi(f(x))} \otimes \mathcal{L}_\chi)$$

of the form

$$\sum_{s_1, ..., s_{a+b} \in E, s_{a+b} \ne 0} \chi(s_{a+b}) \psi(A_a(-1)^{a+1} a s_a + A_{-b}(-1)^{b+1} b s_{-b} + R)$$

with

$$R := P(s_1, ..., s_{a-1}) + Q(s_{-1}, ..., s_{-(b-1)}).$$

Making use of the relations $s_{-i} = s_{a+b-i}/s_{a+b}$, we see that the term $Q(s_{-1}, ..., s_{-(b-1)})$ is itelf a polynomial in $s_{a+b-1}, s_{a+b-2}, ..., s_{a+1}$ and $1/s_{a+b}$. Thus R does not involve the variable s_a. Using the relation $s_{-b} = s_a/s_{a+b}$, we see that the only occurrence of the variable s_a is in the two terms

$$A_a(-1)^{a+1} a s_a + A_{-b}(-1)^{b+1} b s_a/s_{a+b}.$$

So we can factor out the sum over s_a, and get an expression of $\det(-Frob_E)$ as the product of

$$\sum_{s_1, .., s_{a-1}, s_{a+1}, ..., s_{a+b} \in E, s_{a+b} \ne 0} \chi(s_{a+b}) \psi(P(s_1, ..., s_{a-1}) + Q(s_{a+1}, ..., s_{a+b-1}, 1/s_{a+b}))$$

times

$$\sum_{s_a} \psi(s_a(A_a(-1)^{a+1} a + A_{-b}(-1)^{b+1} b/s_{a+b})).$$

This last sum vanishes unless

$$s_{a+b} = A_{-b}(-1)^{b+1} b/A_a(-1)^a a,$$

in which case this sum is equal to $\#E$. Defining $\alpha = A_{-b}(-1)^{b+1}b/A_a(-1)^a a$ we get an expression for $\det(-Frob_E)$ as

$$\chi(\alpha) \times (\text{an expression independent of} \chi).$$

Putting this all together, we get

$$\det(-Frob_{E,\chi}|\omega(N)) = \chi(\alpha)S(E,N)$$

with $S(E,N)$ the exponential sum in $a+b-2$ variables $s_1, ..., s_{a-1}, s_{a+1}, ..., s_a$ given by

$$\#E^{1-(a+b)/2} \sum_{s_1,...,s_{a-1},s_{a+1},...,s_{a+b-1}\in E} \psi(P(s_1,...,s_{a-1})+Q(s_{a+1},...,s_{a+b-1},1/\alpha)$$

Taking $\chi = \mathbb{1}$, we see that

$$S(E,N) = \det(-Frob_{E,\mathbb{1}}|\omega(N)).$$

Thus we find

$$\det(Frob_{E,\chi}|\omega(N)) = \chi(\alpha)\det(Frob_{E,\mathbb{1}}|\omega(N)),$$

so defining

$$\beta := \det(Frob_{k,\mathbb{1}}|\omega(N)),$$

we get the arithmetic determinant formula

$$\text{``}\det\text{''}(N) = \beta^{\deg} \otimes \delta_\alpha.$$

$\square$

The following lemma gives us some control over the quantity $\beta := \det(Frob_{k,\mathbb{1}}|\omega(N))$.

Lemma 21.2. *Let k be a finite field of characteristic p, ψ a nontrivial additive character of k. Let $f(x) = \sum_{i=-b}^{a} A_i x^i \in k[x, 1/x]$ be a Laurent polynomial of "bidegree" (a,b), with a, b both ≥ 1 and both prime to p. The ratio*

$$\det(Frob_k|H_c^1(\mathbb{G}_m/\overline{k}, \mathcal{L}_{\psi(f(x))}))^2/(\#k)^{a+b}$$

is a root of unity of order dividing $2p$ if p is odd, and the ratio is ± 1 if $p = 2$. If $a + b$ is even, the same is true for the ratio

$$\det(Frob_k|H_c^1(\mathbb{G}_m/\overline{k}, \mathcal{L}_{\psi(f(x))}))/(\#k)^{(a+b)/2}.$$

If $a + b$ is odd, the same is true of the ratio

$$\det(Frob_k|H_c^1(\mathbb{G}_m/\overline{k}, \mathcal{L}_{\psi(f(x))}))/(G(\psi, \chi_2)(\#k)^{(a+b-1)/2}),$$

with $G(\psi, \chi_2)$ the quadratic Gauss sum.

Proof. The exponential sum expression for $\det(Frob_k|H_c^1(\mathbb{G}_m/\overline{k}, \mathcal{L}_{\psi(f(x))}))$ derived in the proof of the previous theorem shows that this determinant lies in the ring $\mathbb{Z}[\zeta_p]$ of integers in the cyclotomic field $\mathbb{Q}(\zeta_p)$. Hence the ratios in question all lie in $\mathbb{Q}(\zeta_p)$ as well. We are asserting that the ratios in question are each roots of unity in $\mathbb{Q}(\zeta_p)$. For this, it suffices to see that each ratio has absolute value one at every place, finite or infinite, of $\mathbb{Q}(\zeta_p)$. At archimedean places, this is the fact that $H_c^1(\mathbb{G}_m/\overline{k}, \mathcal{L}_{\psi(f(x))})$ is pure of weight one, and of dimension $a + b$. At ℓ-adic places, $\ell \neq p$, both the determinant, calculated now via ℓ-adic cohomology, and the quantities $\#k$ and $G(\psi, \chi_2)$, are ℓ-adic units. At the unique place lying over p, both this determinant and its complex conjugate have the same p-adic absolute value as each other, as the two quantities are Galois conjugate in $\mathbb{Q}_p(\zeta_p)$. But the two groups $H_c^1(\mathbb{G}_m/\overline{k}, \mathcal{L}_{\psi(f(x))})$ and $H_c^1(\mathbb{G}_m/\overline{k}, \mathcal{L}_{\overline{\psi}(f(x))})$ are dually paired to $\overline{\mathbb{Q}_\ell}(-1)$, so the product of the determinant with its complex conjugate is $(\#k)^{a+b}$. Therefore the square of the determinant, divided by $(\#k)^{a+b}$, is a p-adic unit as well. The square of either of the second two ratios is ± 1 times the first ratio, so it too is a p-adic unit. $\square$

Theorem 21.3. *We have the following results concerning the object N of Theorem 21.1.*

(1) *The quantity*

$$\beta := \det(Frob_{k,\mathbb{1}}|\omega(N))$$

is a root of unity of order dividing $2p$, if either $a + b$ is even or if p is odd and -1 is a square in k. If p and $a + b$ are both odd and -1 is a nonsquare in k, then β is a root of unity of order dividing $4p$.

(2) *If $(-1)^a a A_a = (-1)^{b+1} b A_{-b}$, then for any $a + b$'th root γ of $1/\beta$, the object $\gamma^{deg} \otimes N$ has $G_{geom} \subset G_{arith} \subset SL(a + b)$.*

(3) *If in addition $a \neq b$ and $\gcd\{i|A_i \neq 0\} = 1$ and $a+b$ is a prime number ≥ 3, then the object $\gamma^{deg} \otimes N$ has $G_{geom} = G_{arith} = SL(a + b)$.*

(4) *If the object $\gamma^{deg} \otimes N$ has $G_{geom} = G_{arith} = SL(a+b)$ (e.g., if $a+b$ is prime), then for any $r \geq 2$ distinct characters $\chi_1, ..., \chi_r$ of $k^\times$, the object $\oplus_{i=1}^r \gamma^{deg} \otimes N \otimes \mathcal{L}_{\chi_i}$ has*

$$G_{geom} = G_{arith} = \prod_{i=1}^r SL(a + b).$$

Proof. Assertion (1) is immediate from the corollary, because $G(\psi, \chi_2)^2 = \chi_2(-1)\#k$. Assertion (2) is immediate from assertion (2) of Theorem

21.1. To prove (3), we argue as follows. By parts (3) and (4) of Theorem 21.1, G^0_{geom} is an irreducible subgroup of $SL(a+b)$ which lies in no orthogonal subgroup $SO(a+b)$. By Gabber's theorem on prime-dimensional representations [**Ka-ESDE**, 1.6], the only connected irreducible subgroups of $SL(a+b)$, for $a+b$ prime, are $SL(a+b)$, $SO(a+b)$, the image of $SL(2)$ in $\mathrm{Sym}^{a+b-1}(std_2)$, and, if $a+b=7$, the group G_2 in its seven-dimensional representation. All of these except for $SL(a+b)$ itself lie in an orthogonal subgroup $SO(a+b)$.

Assertion (4) results from the Goursat-Kolchin-Ribet Theorem 13.5. We have a priori inclusions for this object

$$G_{geom} \subset G_{arith} \subset \prod_{i=1}^{r} SL(a+b),$$

so it suffices to show that $G_{geom} = \prod_{i=1}^{r} SL(a+b)$. We must show that for $i \neq j$, and for any one-dimensional object $L \in \mathcal{P}_{geom}$, there is no geometric isomorphism between $N \otimes \mathcal{L}_{\chi_i}$ and either $N \otimes \mathcal{L}_{\chi_j} \star_{mid} L$ or its Tannakian dual. To see this, we argue as follows. The fact that $N \otimes \mathcal{L}_{\chi_i}$ and $N \otimes \mathcal{L}_{\chi_j}$ and its Tannakian dual all have trivial determinants forces $L^{\otimes(a+b)}$ to be geometrically trivial, which in turn implies that L is punctual, so some δ_ϵ with $\epsilon \in \overline{k}^{\times}$. Thus we must show that $\mathcal{L}_{\psi(f(x))} \otimes \mathcal{L}_{\chi_i}$ is not geometrically isomorphic to either $\mathcal{L}_{\psi(f(\epsilon x))} \otimes \mathcal{L}_{\chi_j}$ or to $\mathcal{L}_{\psi(-f(\epsilon/x))} \otimes \mathcal{L}_{\chi_j}$, for any ϵ. The second is impossible, because of the asymmetry of the Swan conductors. For the first, their ratio is $\mathcal{L}_{\psi(f(\epsilon x)-f(x))} \otimes \mathcal{L}_{\chi_j/\chi_i}$. If $\epsilon \neq 1$, this ratio is wildly ramified at either 0 or ∞ or both, thanks to the hypothesis that f is Artin-Schreier reduced and has $gcd\{i|A_i \neq 0\} = 1$. So their ratio is not geometrically trivial. If $\epsilon = 1$, their ratio is $\mathcal{L}_{\chi_j/\chi_i}$, which is nontrivially ramified at both 0 and ∞, so again is not geometrically trivial. $\square$

To end this chapter, we give another family of examples where, for every $n \geq 3$ we have $G_{geom} \subset G_{arith} \subset SL(n)$, but where, once again, only when n is prime can we prove that $G_{geom} = G_{arith} = SL(n)$. We work over a finite field k, with ψ a nontrivial additive character of k. For each integer $m \geq 1$, we denote by Kl_m the Kloosterman sheaf of rank m $Kl_m(\psi; \mathbb{1}, \mathbb{1}, ..., \mathbb{1})$. Recall that its complex conjugate $\overline{Kl_m}$ is given by $\overline{Kl_m} = [x \mapsto (-1)^m x]^\star Kl_m$.

Theorem 21.4. *Fix strictly positive integers $a \neq b$. Denote by $\mathcal{F}$ the lisse sheaf on $\mathbb{G}_m$ given by*

$$\mathcal{F} := Kl_a \otimes [x \mapsto 1/x]^\star \overline{Kl_b} = Kl_a \otimes [x \mapsto (-1)^b/x]^\star Kl_b.$$

Denote by $N \in \mathcal{P}_{arith}$ the object $\mathcal{F}((a+b-1)/2)[1]$. Then we have the following results.

(1) The object $N \in \mathcal{P}_{arith}$ is pure of weight zero, geometrically irreducible, and of "dimension" $a + b$.
(2) For a multiplicative character χ of $k^\times$, consider the Kloosterman sheaf $Kl_{a+b}(\psi; \chi, .., \chi, \mathbb{1}, ..., \mathbb{1})$ of rank $a + b$ with characters χ repeated a times, and $\mathbb{1}$ repeated b times. Then we have the identity

$$\det(1 - TFrob_{k,\chi}|\omega(N))$$
$$= \det(1 - TFrob_{k,(-1)^b}|Kl_{a+b}(\psi; \chi, .., \chi, \mathbb{1}, ..., \mathbb{1})((a + b - 1)/2)).$$

(3) We have $G_{geom,N} \subset G_{arith,N} \subset SL(a + b)$.
(4) The object $N \in \mathcal{P}_{arith}$ is geometrically Lie-irreducible, i.e., $G^0_{geom,N}$ is an irreducible subgroup of $SL(a + b)$.
(5) The object $N \in \mathcal{P}_{arith}$ is not geometrically Lie-self-dual, i.e., $G^0_{geom,N}$ lies in no orthogonal group $SO(a + b)$ and, if $a + b$ is even, it lies in no symplectic group $Sp(a + b)$.
(6) If $a + b$ is a prime number, then $G_{geom} = G_{arith} = SL(a + b)$.
(7) Suppose that $G_{geom} = G_{arith} = SL(a+b)$, and that $\gcd(a, b) = 1$ (e.g., both hold if $a + b$ is prime). For any $r \geq 2$ distinct characters $\chi_1, ..., \chi_r$ of $k^\times$, put $N_i := N \otimes \mathcal{L}_{\chi_i}$. Then the object $\oplus^r_{i=1}N_i$ has

$$G_{geom,\oplus^r_{i=1}N_i} = G_{arith,\oplus^r_{i=1}N_i} = \prod_{i=1}^r SL(a + b).$$

Proof. (1) One knows that Kl_m is pure of weight $m - 1$, so $\mathcal{F}$ is pure of weight $a + b - 2$, and hence N is pure of weight zero. To show that N, or equivalently $\mathcal{F}$, is geometrically irreducible, we repeat the argument given in Remark 18.14. By purity, $\mathcal{F}$ is geometrically semisimple, so it suffices to observe that its local monodromy at 0 (or at ∞, either one will do) is indecomposable. Its local monodromy at 0 is $Unip(a) \otimes Wild_{b,1}$, where $Wild_{b,1}$ is totally wild of rank b and Swan conductor one, and the indecomposability follows from [**Ka-RLS**, 3.1.7]. Its local monodromy at ∞ is $Wild_{a,1} \otimes Unip(b)$. Thus the dimension of N is $Swan_0 + Swan_\infty = a + b$.

To prove (2), it suffices to prove, over all finite extensions of k, the identity of traces

$$\mathrm{Trace}(Frob_{k,\chi}|\omega(N)) = \mathrm{Trace}(Frob_{k,(-1)^b}|Kl_{a+b}(\psi; \chi, .., \chi, \mathbb{1}, ..., \mathbb{1})((a+b-1)/2)$$

Because $\mathcal{F}$ is totally wild at both 0 and ∞, there are no bad characters, so we have

$$\mathrm{Trace}(Frob_{k,\chi}|\omega(N)) := \mathrm{Trace}(Frob_k|H^0_c(\mathbb{G}_m \otimes \overline{k}, N \otimes \mathcal{L}_\chi))$$

$$= \mathrm{Trace}(Frob_k|H_c^1(\mathbb{G}_m \otimes \overline{k}, \mathcal{F} \otimes \mathcal{L}_\chi)((a+b-1)/2)).$$

So what we must show is the identity

$$\mathrm{Trace}(Frob_k|H_c^1(\mathbb{G}_m \otimes \overline{k}, \mathcal{F} \otimes \mathcal{L}_\chi))$$

$$= \mathrm{Trace}(Frob_{k,(-1)^b}|Kl_{a+b}(\psi; \chi, .., \chi, \mathbb{1}, ..., \mathbb{1})).$$

On the left-hand side, the H_c^1 is the only nonvanishing cohomology group, so by the Lefschetz trace formula [**Gr-Rat**] the left-hand side is

$$- \sum_{x \in k^\times} \chi(x)\mathrm{Trace}(Frob_{k,x}|Kl_a)\mathrm{Trace}(Frob_{k,(-1)^b/x}|Kl_b).$$

This in turn is

$$- \sum_{x \in k^\times} \mathrm{Trace}(Frob_{k,x}|Kl_a(\psi; \chi, ..., \chi))\mathrm{Trace}(Frob_{k,(-1)^b/x}|Kl_b),$$

which is precisely the trace of $Frob_{k,(-1)^b}$ on the convolution of $Kl_a(\psi; \chi, ..., \chi)$ with Kl_b, and that convolution is precisely the Kloosterman sheaf $|Kl_{a+b}(\psi; \chi, .., \chi, \mathbb{1}, ..., \mathbb{1})$, cf. [**Ka-GKM**, 5.5].

To prove (3), it suffices to prove, thanks to (2), that for every χ we have the identity

$$\det(Frob_{k,(-1)^b}|Kl_{a+b}(\psi; \chi, .., \chi, \mathbb{1}, ..., \mathbb{1})) = (\#k)^{(a+b)(a+b-1)/2}.$$

This identity is a special case of the arithmetic determinant formula for Kloosterman sheaves given in [**Ka-GKM**, 7.4.1.3 and 7.4.1.4]. [There the formula comes out as $(\#k)^{(a+b)(a+b-1)/2}$ times the additional factor $\chi^a((-1)^{a+b-1})\chi^a((-1)^b)$, but this factor is 1.]

To prove (4), we must show that N is not geometrically isomorphic to any nontrivial multiplicative translate of itself. But already this is true for its local monodromy at 0, $Unip(a) \otimes Wild_{b,1}$, indeed it is true for $Wild_{b,1}$ itself, cf. [**Ka-GKM**, 4.1.6 (3)].

We prove (5) exactly as in Theorem 21.1. We simply observe that N has different Swan conductors a and b at 0 and ∞ respectively, as do all its direct images by Kummer maps.

We then get (6) by Gabber's classification of prime-dimensional representations [**Ka-ESDE**, 1.6], exactly as in the proof of Theorem 21.3.

To prove (7), we use Goursat-Kolchin-Ribet. We have a priori inclusions for the object $\oplus_{i=1}^r N_i$,

$$G_{geom, \oplus_{i=1}^r N_i} \subset G_{arith, \oplus_{i=1}^r N_i} \subset \prod_{i=1}^r SL(a+b),$$

so it suffices to show that $G_{geom, \oplus_{i=1}^r N_i} = \prod_{i=1}^r SL(a+b)$. We must show that for $i \neq j$, and for any one-dimensional object $L \in \mathcal{P}_{geom}$, there is no geometric isomorphism between $N \otimes \mathcal{L}_{\chi_i}$ and either $(N \otimes \mathcal{L}_{\chi_j}) \star_{mid} L$ or its Tannakian dual. To see this, we argue as follows. The fact that $N \otimes \mathcal{L}_{\chi_i}$ and $N \otimes \mathcal{L}_{\chi_j}$ and its Tannakian dual all have trivial determinants forces the Tannakian tensor power $L^{\otimes(a+b)}$ to be geometrically trivial, which in turn implies that L is punctual, so some δ_ϵ with $\epsilon \in \overline{k}^\times$. Thus we must show that for $i \neq j$, $N \otimes \mathcal{L}_{\chi_i}$ is not geomerically isomorphic to any multiplicative translate of either $N \otimes \mathcal{L}_{\chi_j}$ or of its Tannakian dual. The second is impossible, because of the asymmetry of Swan conductors. To rule out the first, we argue as follows. If $N \otimes \mathcal{L}_{\chi_i}$ is geometrically isomorphic to some multiplicative translate $[x \mapsto \epsilon x]^\star (N \otimes \mathcal{L}_{\chi_j})$, then comparing local monodromies at 0 and ∞ respectively we find geometric isomorphisms

$$Wild_{b,1} \otimes \mathcal{L}_{\chi_i} \cong [x \mapsto \epsilon x]^\star (Wild_{b,1} \otimes \mathcal{L}_{\chi_j})$$

and

$$Wild_{a,1} \otimes \mathcal{L}_{\chi_i} \cong [x \mapsto \epsilon x]^\star (Wild_{a,1} \otimes \mathcal{L}_{\chi_j}).$$

Suppose first that $a = 1$. Then $Wild_{a,1} \cong \mathcal{L}_{\psi(x)}$ as $I(\infty)$ representation, so the second isomorphism asserts that $\mathcal{L}_{\psi(x)} \otimes \mathcal{L}_{\chi_i} \cong \mathcal{L}_{\psi(\epsilon x)} \otimes \mathcal{L}_{\chi_j}$ as $I(\infty)$ representation, i.e., that $\mathcal{L}_{\psi((1-\epsilon)x)} \otimes \mathcal{L}_{\chi_i/\chi_j}$ is the trivial character of $I(\infty)$. This is nonsense: if $\epsilon \neq 1$, this character has Swan conductor one, and if $\epsilon = 1$, it is the nontrivial character $\mathcal{L}_{\chi_i/\chi_j}$. Similarly, we deal with the case $b = 1$.

If both $a, b \geq 2$, then both $Wild_{a,1}$ and $Wild_{b,1}$ have trivial determinants. Equating determinants in the two isomorphisms, we find equalities $\chi_i^b = \chi_j^b$ and $\chi_i^a = \chi_j^a$. As $gcd(a,b) = 1$, we infer that $\chi_i = \chi_j$, contradiction. $\square$

CHAPTER 22

$SL(n)$ **Examples with Slightly Composite** n

In this chapter, we continue to study the object N of Theorem 21.1. Thus k is a finite field of characteristic p, ψ a nontrivial additive character of k, $f(x) = \sum_{i=-b}^{a} A_i x^i \in k[x, 1/x]$ is a Laurent polynomial of "bidegree" (a, b), with a, b both ≥ 1 and both prime to p. We assume that $f(x)$ is Artin-Schreier reduced. We take for N the object $N := \mathcal{L}_{\psi(f(x))}(1/2)[1] \in \mathcal{P}_{arith}$.

Theorem 22.1. *The object N, viewed in $<N>_{geom}$, is not geometrically isomorphic to the middle convolution of any two objects K and L in $<N>_{geom}$ each of which has dimension ≥ 2. Equivalently, the representation of $G_{geom,N}$ corresponding to N is not the tensor product of two other representations of $G_{geom,N}$, each of which has dimension ≥ 2.*

Proof. The key point is that the object N has no bad characters (because it is totally wildly ramified at both 0 and ∞, a property equivalent to having no bad characters). Therefore every object in $<N>_{arith}$ shares this property of having no bad characters, cf. Theorem 4.1. In other words, every object $M \in < N >_{arith}$ is (strictly speaking, its $\mathcal{H}^{-1}(M)$ is) totally wildly ramified at both 0 and ∞. But every object in $<N>_{geom}$ is, geometrically, a direct summand of some object of $<N>_{arith}$ (indeed, of some multiple middle convolution of N and of its Tannakian dual $N^\vee$), so itself is totally wildly ramified at both 0 and ∞. But N has generic rank one, so the theorem results from the following theorem. $\qquad\square$

Theorem 22.2. *Let K and M be two irreducible objects of $\mathcal{P}_{geom}$, each of which is nonpunctual and totally wildly ramified at both 0 and ∞. Then their middle convolution $K \star_{mid} M$ has generic rank ≥ 2.*

Proof. We have $K = \mathcal{K}[1]$ and $M = \mathcal{M}[1]$ for irreducible middle extension sheaves $\mathcal{K}$ and $\mathcal{M}$ on $\mathbb{G}_m/\overline{k}$, each of which is totally wildly ramified at both 0 and ∞. Over a dense open set U, their middle convolution is of the form $\mathcal{Q}[1]$, for $\mathcal{Q}$ a lisse sheaf on U whose stalk at a point $a \in U(\overline{k})$ is the image of the "forget supports" map

$$H^1_c(\mathbb{G}_m/\overline{k}, \mathcal{K} \otimes [x \mapsto a/x]^\star \mathcal{M}) \to H^1(\mathbb{G}_m/\overline{k}, \mathcal{K} \otimes [x \mapsto a/x]^\star \mathcal{M}).$$

We claim that, because $\mathcal{K}$ and $\mathcal{M}$ are totally wildly ramified at both 0 and ∞, the tensor product sheaf $\mathcal{K} \otimes [x \mapsto a/x]^\star \mathcal{M}$ is, for all but at most finitely many a, itself totally wildly ramified at both 0 and ∞. Let us temporarily grant this claim. Then for good a, the forget supports map is an isomorphism, and hence the generic rank of $\mathcal{Q}$ is

$$-\chi_c(\mathbb{G}_m/\overline{k}, \mathcal{K} \otimes [x \mapsto a/x]^\star \mathcal{M})$$

$$= Swan_0 + Swan_\infty + \sum_{b \in \mathbb{G}_m(\overline{k})} (drop_b + Swan_b)$$

$$\geq Swan_0 + Swan_\infty \geq 1 + 1,$$

the last inequality because $\mathcal{K} \otimes [x \mapsto a/x]^\star \mathcal{M}$ is totally wildly ramified at both 0 and ∞. Thus we are reduced to the following lemma, applied to the $I(0)$ (resp. the $I(\infty)$) representations of $\mathcal{K}$ and of $[x \mapsto 1/x]^\star \mathcal{M}$. $\square$

Lemma 22.3. *Let R and S be ℓ-adic representations of $I(0)$ (resp. of $I(\infty)$) which are both totally wildly ramified. Then for all but finitely many $a \in \mathbb{G}_m(\overline{k})$, $R \otimes [x \mapsto ax]^\star S$ is a totally wild ramified representation of $I(0)$ (resp. of $I(\infty)$).*

Proof. We treat the $I(0)$ case; the $I(\infty)$ case is identical. The wildness of an ℓ-adic representation of $I = I(0)$ depends only on its restriction to the wild inertia group $P = P(0)$. This restriction is semisimple with finite image, simply because P is a pro-p group, and $\ell \neq p$. So we may replace R and S by their semisimplifications as I-representations, then reduce to the case where R and S are each I-irreducible.

The next step is to reduce further to the case in which R and S are both P-irreducible. We claim that after some Kummer pullback by an n'th power map, for some n prime to p, both R and S are P-isomorphic to direct sums of irreducible P-representations, each of which extends to an I-representation. [Such a Kummer pullback is harmless for questions of total wildness, as P is unchanged.] Recall that once we pick an element $\gamma \in I$ whose pro-order is prime to p and which mod P is a topological generator of the tame quotient $I^{tame} \cong \widehat{\mathbb{Z}}_{not\ p}(1)$, we have a semidirect product expression

$$I \cong P \rtimes <\gamma>.$$

Since R (resp. S) is I-irreducible, conjugation by γ must permute the finitely many isomorphism classes of irreducible P-representations which occur in it. So replacing γ by a prime to p power of itself, which amounts to passing to a Kummer pullback, we get a situation where each P-irreducible in R (resp. in S) is isomorphic to its γ-conjugate, hence extends to a representation of I (and the extended representation of I is unique up to tensoring with a Kummer sheaf $\mathcal{L}_\chi$, i.e., with a character of $I/P = I^{tame}$).

So we are reduced to the case where R and S remain irreducible when restricted to P. If $R^\vee$ and S are inequivalent as P-representations, then $(R \otimes S)^P = Hom_P(R^\vee, S) = 0$, i.e., $R \otimes S$ is totally wild. So if no multiplicative translate of S is P-isomorphic to $R^\vee$, we are done.

If some multiplicative translate of S is P-isomorphic to $R^\vee$, say S itself, then $(R \otimes S)^P = Hom_P(R^\vee, S)$ is one-dimensional, with trivial P action, so is some Kummer sheaf $\mathcal{L}_\chi$. In this latter case, we have an I-isomorphism

$$Hom_P(R^\vee, S) \otimes R^\vee \cong S,$$

i.e.,

$$S \cong R^\vee \otimes \mathcal{L}_\chi.$$

Replacing R by $R \otimes \mathcal{L}_{\overline{\chi}}$, which is harmless for questions of total wildness, we must treat the case in which $S \cong R^\vee$ as I-representation. In view of the previous paragraph, it suffices to show that there are at most finitely many $a \in \mathbb{G}_m(\overline{k})$ such that S is P-isomorphic to $[x \mapsto ax]^\star S$. In fact, we will show that there are at most $\text{Max}(Swan(S), Swan(\det(S)))$ such translates.

We first recall [**Ka-GKM**, 4.1.6 (2)] that there are at most $Swan(S)$ $a \in \mathbb{G}_m(\overline{k})$ such that S is I-isomorphic to $[x \mapsto ax]^{-1}S$. To see this, we consider the canonical extension, say $\mathcal{S}$, of S to a lisse sheaf on $\mathbb{G}_m$ which is tame at ∞ and whose $I(0)$-representation is isomorphic to S. Because S is I-irreducible, $\mathcal{S}$ is irreducible as lisse sheaf on $\mathbb{G}_m/\overline{k}$. If the I-isomorphism class of S is invariant under multiplicative translation by some $a \neq 1$ in $\mathbb{G}_m(\overline{k})$, say by a of multiplicative order $n > 1$ prime to p, then the isomorphism class of $\mathcal{S}$ is invariant under multiplicative translation by μ_n, and hence $\mathcal{S}$ descends through the n'th power map, i.e., $\mathcal{S} \cong [n]^\star \mathcal{T}$ for some lisse sheaf $\mathcal{T}$ on $\mathbb{G}_m$, in which case $Swan_0(\mathcal{S}) = n \times Swan_0(\mathcal{T})$. In particular, n divides $Swan_0(\mathcal{S}) = Swan(S)$. So there are at most $Swan(S)$ multiplicative translates of S which are I-isomorphic to S.

We next infer from this that there are at most

$$\text{Max}(Swan(S), Swan(\det(S)))$$

$a \in \mathbb{G}_m(\overline{k})$ such that S is P-isomorphic to $[x \mapsto ax]^\star S$. Consider first the case in which $\det(S)$ is itself wildly ramified. Then applying the above result to $\det(S)$, there are at most $Swan(\det(S))$ values $a \in \mathbb{G}_m(\overline{k})$ such that $\det(S)$ is I-isomorphic to $[x \mapsto ax]^\star \det(S)$. But taking a Kummer sheaf $\mathcal{L}_\chi$ which has the same value on γ as $\det(S)$, the ratio $\det(S)/\mathcal{L}_\chi$ is a wild character which takes only p-power roots of unity as values. Indeed, as S has p-power rank (being P-irreducible), we may replace S by its twist by the unique $rank(S)$'th root of $\mathcal{L}_{\overline{\chi}}$, and reduce to the case where $\det(S)$ has p-power order. In this case, $\det(S)$ is a character of $I \cong P \rtimes <\gamma>$ which is trivial on γ, so is completely determined by its restriction to P. So in this case, the P-isomorphism class of $\det(S)$ is preserved by at most $Swan(\det(S))$ multiplicative translations.

Now consider the case in which $\det(S)$ is tame. Here the I-isomorphism class of $\det(S)$ is invariant by multiplicative translation. But one knows that for an irreducible representation S of I which is P-irreducible, its I-isomorphism class is determined by the two data consisting of the P-isomorphism class of S and the I-isomorphism class of $\det(S)$, cf. [**Ka-ClausCar**, 2.5.1]. [Indeed, as noted earlier, if S_1 and S_2 are P-irreducible and P-isomorphic, then for some χ, $S_1 \cong S_2 \otimes \mathcal{L}_\chi$. Taking determinants, we get $\det(S_1) \cong \det(S_2) \otimes \mathcal{L}_{\chi^q}$, for q the common rank of the S_i. But q is some power q of p, by P-irreducibility, so if the determinants are isomorphic then χ is trivial.] Since the I-isomorphism class of S is preserved by at most $Swan(S)$ translations, the same is true for its P-isomorphism class. $\qquad\square$

Theorem 22.4. *We have the following result concerning the object N of Theorem 21.1. Suppose that $a \neq b$, that $\gcd\{i|A_i \neq 0\} = 1$, and that $(-1)^a a A_a = (-1)^{b+1} b A_{-b}$. Suppose further that $a + b$ is the product $\ell_1 \ell_2$ of two distinct primes. Then $G^0_{geom} \subset SL(\ell_1 \ell_2)$, already shown to be a connected irreducible subgroup which is not self-dual, has a simple Lie algebra.*

Proof. We argue by contradiction. If $Lie(G^0_{geom})$ is not simple, then G^0_{geom} is the image in $SL(\ell_1 \ell_2)$ of a product group $G_1 \times G_2$, with $G_1 \subset SL(\ell_1)$ and $G_2 \subset SL(\ell_2)$ connected irreducible subgroups, at least one of which is not self-dual, under the tensor product of the given representations.

Suppose first that $2 = \ell_1$. Then for lack of choice $G_1 = SL(2)$, and hence by Gabber's theorem on prime-dimensional representations [**Ka-ESDE**, 1.6], G_2 must be $SL(\ell_2)$, the only non-self-dual choice. The image of this $G_1 \times G_2$ in $SL(\ell_1 \ell_2)$ is this product group, and it is its own normalizer in $SL(\ell_1 \ell_2)$. Therefore $G_{geom} = G_1 \times G_2$. This contradicts Theorem 22.1.

Suppose now that both ℓ_1 and ℓ_2 are odd primes. Then at least one of the factors, say G_1, is not self-dual, so must be $SL(\ell_1)$. The second factor G_2 is either $SL(\ell_2)$ or $SO(\ell_2)$ or $SO(3)$, viewed as the image of $SL(2)$ in $Sym^{\ell_2-1}(std_2)$, or, if $\ell_2 = 7$, possibly the exceptional group

$G2$ in its seven-dimensional representation. In this case as well, the image of $G_1 \times G_2$ in $SL(\ell_1\ell_2)$ is this product group, and its normalizer in $SL(\ell_1\ell_2)$ is itself, augmented by the scalars $\mu_{\ell_1\ell_2}$ in $SL(\ell_1\ell_2)$. In the case when G_2 is $SL(\ell_2)$, these scalars are already in $G_1 \times G_2$, so $G_1 \times G_2$ is its own normalizer in $SL(\ell_1\ell_2)$. So in this case G_{geom} is the product group $G_1 \times G_2$, and we contradict Theorem 22.1.

If G_2 is $SO(\ell_2)$ or one of its listed subgroups, G_2 contains no non-trivial scalars, so the normalizer of $G_1 \times G_2$ in $SL(\ell_1\ell_2)$ is the product group

$$G_1 \times (\mu_{\ell_2} \times G_2).$$

Thus we have

$$G_1 \times \times G_2 \subset G_{geom} \subset G_1 \times (\mu_{\ell_2} \times G_2).$$

So G_{geom} is either $G_1 \times \times G_2$ or it is $G_1 \times (\mu_{\ell_2} \times G_2)$, in either case contradicting Theorem 22.1. $\square$

Remark 22.5. In view of this result, it is natural to ask the following question: for which pairs of distinct primes $\ell_1 < \ell_2$ is it the case that the only connected irreducible subgroup of $SL(\ell_1\ell_2)$ which is not self-dual, and whose Lie algebra is simple, is the entire group $SL(\ell_1\ell_2)$ itself? Such a subgroup is the image of an irreducible, non-self-dual, $\ell_1\ell_2$-dimensional representation of a simply connected group with simple Lie algebra. We wish to know the cases in which the only such are the standard representation $std_{\ell_1\ell_2}$ of $SL(\ell_1\ell_2)$ and its contragredient. One knows that only the types $A_n, n \geq 2$, $D_{2k+1}, k \geq 2$, and E_6 have irreducible representations which are not self-dual. According to a recent result of Goldstein, Guralnick and Stong [**GGS**, Cor. 1.5], only the A_n, $n \geq 2$, have irreducible representations which are non-self-dual and of dimension equal to the product $\ell_1\ell_2$ of two distinct primes $\ell_1 < \ell_2$. Moreover, the only dimensions $\ell_1\ell_2$ so attained are of the following three types.

(1) Twin primes: $\ell_2 = \ell_1 + 2$. In general, $n(n+2)$ is the dimension of the representation $\omega_1 + (n-1)\omega_2$ of A_2.

(2) Sophie Germain primes: $\ell_2 = 2\ell_1 + 1$. In general, $n(2n+1)$ is the dimension of the representation $\text{Sym}^2(std_{2n})$ of $SL(2n)$, or the dimension of the representation $\Lambda^2(std_{2n+1})$ of $SL(2n+1)$, or the dimension of the representation $\text{Sym}^{2n+1}(std_3)$ of $SL(3)$.

(3) Variant Sophie Germain primes: $\ell_2 = 2\ell_1 - 1$. In general, $n(2n-1)$ is the dimension of the representation $\text{Sym}^2(std_{2n-1})$ of $SL(2n-1)$, or the dimension of the representation $\Lambda^2(std_{2n})$ of $SL(2n)$, or the dimension of the representation $\text{Sym}^{2n}(std_3)$ of $SL(3)$.

Thus we obtain the following theorem.

Theorem 22.6. *We have the following result concerning the object N of Theorem 21.1. Suppose that $a \neq b$, that $\gcd\{i \,|\, A_i \neq 0\} = 1$, and that $(-1)^a a A_a = (-1)^{b+1} b A_{-b}$. Suppose further that $a + b$ is the product $\ell_1 \ell_2$ of two distinct primes $\ell_1 < \ell_2$. Suppose that these are neither twin primes, Sophie Germain primes, or variant Sophie Germain primes. Then $G_{geom} = SL(\ell_1 \ell_2)$.*

Remark 22.7. Thus for example if $\ell_1 \ell_2 < 400$ and (ℓ_1, ℓ_2) is not any of

$$(2,3), (2,5), (3,5), (3,7), (5,7), (5,11), (7,13), (11,13), (11,23), (17,19),$$

then we have $G_{geom} = SL(\ell_1 \ell_2)$ in the theorem above. This can be seen directly, i.e., without invoking [**GGS**, Cor. 1.5], from the tables of Lübeck [**Lu**]. In these tables, the dimensions listed as valid for all large p are also the dimensions of the same representations of the complex group.

CHAPTER 23

Other $SL(n)$ Examples

In this chapter, we fix an integer $n \geq 3$ which is not a power of the characteristic p, and a monic polynomial $f(x) \in k[x]$ of degree n, $f(x) = \sum_{i=0}^{n} A_i x^i$, $A_n = 1$.

Lemma 23.1. *Suppose that f has n distinct roots in $\overline{k}$, all of which are nonzero (i.e., $A_0 \neq 0$). Let χ be a nontrivial character of $k^\times$ with $\chi^n = 1$. Form the object $N := \mathcal{L}_{\chi(f)}(1/2)[1] \in \mathcal{P}_{arith}$. Its Tannakian determinant "$\det$"(N) is geometrically of finite order. It is geometrically isomorphic to δ_a for $a = (-1)^n A_0 = (-1)^n f(0) =$ the product of all the zeroes of f.*

Proof. Let ρ be a nontrivial character of $k^\times$. Then ρ is good for N, and

$$\det(Frob_{k,\rho}|\omega(N)) = (-1)^n \det(-Frob_k|H_c^1(\mathbb{G}_m/\overline{k}, \mathcal{L}_{\chi(f(x))} \otimes \mathcal{L}_{\rho(x)})/(\#k)^{n/2}$$

$$= (-1)^n \epsilon(\mathcal{L}_{\chi(f(x))} \otimes \mathcal{L}_{\rho(x)})/(\#k)^{n/2},$$

with ϵ the global ϵ constant. The rank one sheaves $\mathcal{L}_{\chi(f(x))}$ and $\mathcal{L}_{\rho(x)}$ have disjoint ramification on $\mathbb{P}^1$, so their ϵ constants are related by

$$\epsilon(\mathcal{L}_{\chi(f(x))} \otimes \mathcal{L}_{\rho(x)})\epsilon(\overline{\mathbb{Q}_\ell})\epsilon(\mathcal{L}_{\chi(f(x))})^{-1}\epsilon(\mathcal{L}_{\rho(x)})^{-1}$$

$$= (\prod_{zeroes\ \alpha_i\ of\ f} \rho(\alpha_i)) \times (\chi(f(0))\chi^n(1)) = \rho((-1)^n f(0))\chi(f(0)),$$

cf. [**De-Const**, Cor. 9.5]. On the other hand, $\epsilon(\mathcal{L}_{\rho(x)}) = 1$, because all its cohomology vanishes. So we get a formula of the form

$$\det(Frob_{k,\rho}|\omega(N)) = (-1)^n \epsilon(\mathcal{L}_{\chi(f(x))} \otimes \mathcal{L}_{\rho(x)})/(\#k)^{n/2}$$

$$= \rho((-1)^n f(0)) \times (a\ factor\ independent\ of\ \rho).$$

This formula, applied to all finite extensions E/k and to all nontrivial characters ρ of $E^\times$, proves the assertion. $\square$

For the rest of this chapter, we make the following three assumptions about f.

(1) f has n distinct roots in $\overline{k}$, all of which are nonzero (i.e., $A_0 \neq 0$).

(2) $gcd\{i|A_i \neq 0\} = 1$.

(3) No multiplicative translate $f(\lambda x)$, $\lambda \in \overline{k}^{\times}$, of $f(x)$ is a $\overline{k}^{\times}$-multiple of the palindrome $f^{pal}(x) := x^n f(1/x) = \sum_i A_{n-i} x^i$ of $f(x)$.

Theorem 23.2. *Let χ be a nontrivial character of $k^{\times}$ with $\chi^n = \mathbb{1}$. Form the object $N := \mathcal{L}_{\chi(f)}(1/2)[1] \in \mathcal{P}_{arith}$. We have the following results.*

(1) *The Tannakian determinant "$\det$"(N) is geometrically of finite order. It is geometrically isomorphic to δ_a for $a = (-1)^n A_0 = (-1)^n f(0) =$ the product of all the zeroes of f.*
(2) *We have $G^0_{geom,N} = SL(n)$.*
(3) *If $(-1)^n f(0) = 1$, then $G_{geom,N} = SL(n)$.*

Proof. Assertion (3) is immediate from (1) and (2). Assertion (1) was proven in Lemma 23.1.

It results from (1) that we have an a priori inclusion $G^0_{geom,N} \subset SL(n)$. From the hypotheses that $gcd\{i | A_i \neq 0\} = 1$ and that $A_0 \neq 0$, it follows that N is not geometrically isomorphic to any nontrivial multiplicative translate of itself, and hence that N is geometrically Lie-irreducible. In other words, $G^0_{geom,N}$ is a connected irreducible subgroup of $SL(n)$.

From the hypothesis that no multiplicative translate $f(\lambda x)$, $\lambda \in \overline{k}^{\times}$, is a $\overline{k}^{\times}$-multiple of the palindrome $f^{pal}(x) := x^n f(1/x)$ of $f(x)$, it follows that no multiplicative translate of N is geometrically isomorphic to its Tannakian dual $N^{\vee} = \mathcal{L}_{\overline{\chi}(f(1/x))}(1/2)[1]$. Indeed, their sets of finite singularites are always different: the singularities of a multiplicative translate of N are the zeros of a multiplicative translate of f, while the singularities of $N^{\vee}$ are the zeroes of $f^{pal}(x) = \sum_i A_{n-i} x^i$. This implies that N is geometrically Lie-non-self-dual, i.e., that $G^0_{geom,N}$ is a connected irreducible subgroup of $SL(n)$ which is not self-dual. [Indeed, if N were geometrically Lie-self-dual, then for some prime-to-p integer $d \geq 1$, $[d]_*(N)$ would be geometrically isomorphic to its dual, namely $[d]_*(N^{\vee})$. Pulling back by $[d]$, we would get a geometric isomorphism $[d]^*[d]_*(N) \cong [d]^*[d]_*(N^{\vee})$. But

$$[d]^*[d]_*(N) \cong \bigoplus_{\zeta \in \mu_d}[x \mapsto \zeta x]^*(N), \quad [d]^*[d]_*(N^{\vee}) \cong \bigoplus_{\zeta \in \mu_d}[x \mapsto \zeta x]^*(N^{\vee}).$$

Matching irreducible constituents, the term $N^{\vee}$ on the right must be geometrically isomorphic to one of the terms $[x \mapsto \zeta x]^*(N)$ on the left.]

Finally we use the fact that $\mathcal{L}_{\chi(f(x))}$ is lisse at both 0 and ∞, and of generic rank one. So the Frobenius torus atttached to $Frob^{ss}_{k,1}$ gives us a torus $Diag(x, 1, ..., 1, 1/x)$ in $G^0_{geom,N}$. We then apply once again the theorem of Kostant and Zarhin [**Ka-ESDE**, 1.2] that the only connected irreducible subgroups of $SL(n)$ containing such a torus are $SO(n)$, $Sp(n)$ if n is even, and $SL(n)$. Since our $G^0_{geom,N}$ is not self-dual, it must be $SL(n)$.   □

Theorem 23.3. *Form the object* $N := \mathcal{L}_{\chi(f)}(1/2)[1] \in \mathcal{P}_{arith}$ *of the previous theorem. Suppose further that* $(-1)^n f(0) = 1$, *so that* $G_{geom,N} = SL(n)$, *and hence for some* $\beta \in \overline{\mathbb{Q}_\ell}^\times$, *the object* $N_0 := N \otimes \beta^{deg}$ *has* $G_{geom,N_0} = G_{arith,N_0} = SL(n)$. *Suppose given* $r \geq 2$ *distinct characters* ρ_i *of* $k^\times$. *Denote by* N_i *the object* $N_0 \otimes \mathcal{L}_{\rho_i}$. *Then the object* $\oplus_i N_i$ *has*

$$G_{geom,\oplus_i N_i} = G_{arith,\oplus_i N_i} = \prod_i SL(n).$$

Proof. In view of the a priori inclusions

$$G_{geom,\oplus_i N_i} \subset G_{arith,\oplus_i N_i} \subset \prod_i SL(n),$$

it suffices to show that $G_{geom,\oplus_i N_i} = \prod_i SL(n)$. By Goursat-Kolchin-Ribet, we must show that for $i \neq j$, N_i is not geometrically isomorphic to either $N_j \star_{mid} L$ or to $N_j^\vee \star_{mid} L$ for any one-dimensional object L. For such an isomorphism to exist, L must be of finite order dividing n, because both N_i and N_j have trivial determinant. Hence L must be punctual, some δ_a. So we must show that N_i is not geometrically isomorphic to any multiplicative translate of either N_j or $N_j^\vee$. But the local monodromy at 0 of N_i is $\mathcal{L}_{\rho_i}$, while the local monodromy at 0 of any multiplicative translate of either N_j or $N_j^\vee$ is $\mathcal{L}_{\rho_j}$.   □

Remark 23.4. Given a monic polynomial $f(x) = \sum A_i x^i \in k[x]$ of degree $n \geq 3$ with $f(0) \neq 0$, which is not a polynomial in x^d for any divisor $d \geq 2$ of n, how can we decide if no multiplicative translate of $f(x)$ is a $\overline{k}^\times$-multiple of its palindrome $f^{pal}(x)$? [We must exclude $n = 2$, because for $f(x) = x^2 + ax + b$ with $b \neq 0$, we have $f(bx) = bf^{pal}(x)$.] For $n \geq 3$, an obvious sufficient condition is that f be "monomially nonpalindromic," i.e., there exists some index i with $0 < i < n$ such that $A_i \neq 0$ but $A_{n-i} = 0$, for then different monomials occur in f^{pal} than in any multiplicative translate of f. For $n \geq 3$, such monomially nonpalindromic f exist, e.g., $x^n + x + 1$. In general, we have the following generic result, whose statement is due to Antonio Rojas-León.

Lemma 23.5. *In the space* $\mathbb{A}^{n-1} \times \mathbb{G}_m \times \mathbb{G}_m$ *of polynomials of degree* $n \geq 3$ *with* $A_0 A_n$ *invertible, the condition that no multiplicative translate of* $f(x)$ *be proportional to* $f^{pal}(x)$ *defines an open dense set.*

Proof. We have already seen that this set is nonempty, as it contains $x^n + x + 1$. So it suffices to see that its complement is closed. The idea is simple. If the zeroes of f are $\alpha_1, ..., \alpha_n$, then the zeroes of f^{pal} are their inverses $1/\alpha_1, ..., 1/\alpha_n$. Suppose there is some λ such that $f(\lambda x)$ has zeroes $1/\alpha_1, ..., 1/\alpha_n$. The zeroes of $f(\lambda x)$ are the α_i/λ, so these are a permutation of the $1/\alpha_i$. In particular, $\alpha_1/\lambda = 1/\alpha_j$ for some j. Thus λ is among the roots of the degree n^2 polynomial $\prod_{i,j=1...,n}(x - \alpha_i\alpha_j)$.

Now in general, given two polynomials $f(x) = \sum_{i=0}^n A_i x^i$ and $g(x) = \sum_{i=0}^n B_i x^i$ with $A_n B_n$ invertible, having roots α_i's and β_j's respectively, the degree n^2 polynomial

$$\{f, g\} := \prod_{i,j=1...,n} (x - \alpha_i\beta_j)$$

has coefficients which are universal polynomials in the A_i/A_n and the B_j/B_n. This is most easily seen by thinking of the tensor product of the corresponding "companion matrices." In other words, view $(1/A_n)f$ (resp. $(1/B_n)g$) as the characteristic polynomial of multiplication by x on $V_f := k[x]/(f(x))$ (resp. of multiplication by y on $V_g := k[y]/(g(y))$). Then $\{f, g\}$ is the characteristic polynomial of their tensor product, multiplication by xy on $V_f \otimes_k V_g \cong k[x, y]/(f(x), g(y))$.

If some multiplicative translate of $f(x)$ is proportional to $f^{pal}(x)$, the factor of proportionality must be A_0/A_n, as one sees by comparing constant terms, and the translate must be by a zero of $\{f, f\}$. So the condition that some multiplicative translate of $f(x)$ be proportional to $f^{pal}(x)$ is that the polynomial

$$\prod_{\text{roots } \lambda \text{ of } \{f,f\}} (f(\lambda x) - (A_0/A_n)f^{pal}(x))$$

vanish identically. As $\{f, f\}$ is monic, with coefficients which are polynomials in the A_i/A_n, this displayed product polynomial itself has coefficients in this same ring, so its vanishing identically is defined by the vanishing of its n^2 coefficients, which are each regular functions on our space $\mathbb{A}^{n-1} \times \mathbb{G}_m \times \mathbb{G}_m$. $\square$

Corollary 23.6. *In the space $\mathbb{A}^{n-1} \times \mathbb{G}_m$ of monic polynomials of degree $n \geq 3$ with A_0 invertible and $A_n = 1$, the condition that no multiplicative translate of $f(x)$ be proportional to $f^{pal}(x)$ defines an open dense set.*

Proof. Just as in the proof of Lemma 23.5, the set is nonempty, because it contains $x^n + x + 1$. Its complement is closed, being the intersection with $A_n = 1$ of the closed complement in the proof of Lemma 23.5. $\square$

CHAPTER 24

An $O(2n)$ Example

In this chapter, we work over a finite field k of odd characteristic. Fix an even integer $2n \geq 4$ and a monic polynomial $f(x) \in k[x]$ of degree $2n$, $f(x) = \sum_{i=0}^{2n} A_i x^i$, $A_{2n} = 1$. We make the following three assumptions about f.

(1) f has $2n$ distinct roots in $\overline{k}$, and $A_0 = -1$.
(2) $\gcd\{i | A_i \neq 0\} = 1$.
(3) f is antipalindromic, i.e., for $f^{pal}(x) := x^{2n} f(1/x)$, we have $f^{pal}(x) = -f(x)$.

Theorem 24.1. *For χ_2 the quadratic character of $k^\times$, form the object $N := \mathcal{L}_{\chi_2(f)}(1/2)[1] \in \mathcal{P}_{arith}$. We have the following results.*

(1) $G_{geom,N} = O(2n)$.
(2) *Choose a nontrivial additive character ψ of k, and define*

$$\alpha := (-1/\sqrt{\#k}) \sum_{x \in k^\times} \psi(x) \chi_2(x),$$

the normalized Gauss sum. Then for the constant-field twisted object $N_\alpha := \alpha^{\deg} \otimes N$ we have

$$G_{geom,N_\alpha} = G_{arith,N_\alpha} = O(2n).$$

(3) *If -1 is a square in k, then $G_{geom,N} = G_{arith,N} = O(2n)$.*

Proof. (3) is a special case of (2), simply because $\alpha = \pm 1$ when -1 is a square in k. And (1) results from (2), since N and N_α are geometrically isomorphic. To prove (2), we argue as follows. The complex conjugate of α is $\chi_2(-1)\alpha$. This, together with the fact that f is antipalindromic, shows that N_α is arithmetically self-dual. As N_α is irreducible, the duality is either symplectic or orthogonal. So either we have

$$G_{geom,N_\alpha} \subset G_{arith,N_\alpha} \subset O(2n)$$

or we have

$$G_{geom,N_\alpha} \subset G_{arith,N_\alpha} \subset Sp(2n).$$

The second case is impossible, because by Lemma 23.1, the determinant of N (and so also the determinant of N_α) is geometrically δ_{-1}, so we cannot have $G_{geom,N_\alpha} \subset Sp(2n)$. Therefore the duality is orthogonal. Just as in the proof of Theorem 23.2, hypothesis (2) on f implies that N_α is geometrically Lie-irreducible. The Frobenius torus argument of Theorem 23.2 then shows that G^0_{geom,N_α} must be $SO(2n)$. But the determinant of N_α is geometrically nontrivial, so from the inclusions

$$SO(2n) = G^0_{geom,N_\alpha} \subset G_{geom,N_\alpha} \subset G_{arith,N_\alpha} \subset O(2n)$$

we infer (2). □

Remark 24.2. A monic antipalindromic polynomial of degree $2n$ over a field k of odd characteristic can be written uniquely as $x^n(P(x) - P(1/x))$ with $P(x)$ a monic polynomial of degree n with vanishing constant term. In this way, the space $MonicAntipal_{2n}$ of monic antipalindromic polynomials of degree $2n$ becomes the affine space $\mathbb{A}^{n-1}$. In $MonicAntipal_{2n}$, the condition of having $2n$ distinct roots, i.e., of having an invertible discriminant, defines an open dense set. Indeed, the set is obviously open, so it suffices to observe that it is nonempty. If $2n - 1$ is invertible in k, then the polynomial $x^{2n} + x^{2n-1} - x - 1$ has all distinct roots (as it is $(x + 1)(x^{2n-1} - 1)$). If If $2n$ is invertible in k, then $x^{2n} - 1$ has all distinct roots.

CHAPTER 25

G_2 Examples: the Overall Strategy

In this and the next two chapters, we fix, for each prime p, a prime $\ell \neq p$ and a choice of nontrivial $\overline{\mathbb{Q}_\ell}^\times$-valued additive character ψ of the prime field $\mathbb{F}_p$. Given a finite extension field $k/\mathbb{F}_p$, we take as nontrivial additive character of k the composition $\psi_k := \psi \circ \mathrm{Tr}_{k/\mathbb{F}_p}$, whenever a nontrivial additive character of k is (implicitly or explicitly) called for (for instance in the definition of a Kloosterman sheaf, or of a hypergeometric sheaf, on $\mathbb{G}_m/k$). Given a finite field k of characteristic p, and a (possibly trivial) multiplicative character χ of $k^\times$ which, if p is odd, is **not** the quadratic character χ_2, we form the following lisse sheaf $\mathcal{F}(\chi, k)$ on $\mathbb{G}_m/k$. If $p = 2$, we take the Tate-twisted Kloosterman sheaf of rank seven

$$\mathcal{F}(\chi, k) := Kl(\mathbb{1}, \mathbb{1}, \mathbb{1}, \chi, \chi, \overline{\chi}, \overline{\chi})(3).$$

If p is odd, we take the "Gauss sum twisted" hypergeometric sheaf of type $(7, 1)$

$$\mathcal{F}(\chi, k) := (A^{-7})^{deg} \otimes \mathcal{H}(\mathbb{1}, \mathbb{1}, \mathbb{1}, \chi, \chi, \overline{\chi}, \overline{\chi}; \chi_2),$$

with A the negative of the quadratic Gauss sum:

$$A := -\sum_{x \in k^\times} \chi_2(x)\psi_k(x).$$

It is proven in [**Ka-G2Hyper**, 9.1] that each of these lisse sheaves is pure of weight zero, orthogonally self-dual, and has $G_{geom} = G_{arith} = G_2$ (with G_2 seen as a subgroup of $SO(7)$ via G_2's irreducible seven-dimensional representation).

We think of this result in the following way: for $\#k$ large, and χ fixed, the semisimplifications of the $\#k^\times$ Frobenius conjugacy classes $\{Frob_{a,k}|\mathcal{F}(\chi, k)\}_{a \in k^\times}$ attached to the lisse sheaf $\mathcal{F}(\chi, k)$ at the points $a \in k^\times$ are approximately equidistributed in the space of conjugacy classes of the compact group UG_2, the compact form of G_2. What we would like to prove (but cannot, at present) is that, for $\#k$ large, if we **fix** a point $a \in k^\times$, the semisimplifications of the ($\#k^\times$, if $p = 2$, $\#k^\times - 1$ if p is odd) Frobenius conjugacy classes $\{Frob_{a,k}|\mathcal{F}(\chi, k)\}_\chi$

indexed by the characters χ of $k^{\times}$ (with $\chi \neq \chi_2$ when p is odd) are approximately equidistributed in the space of conjugacy classes of UG_2.

Recall that in UG_2, conjugacy classes are determined by their characteristic polynomials in the irreducible seven-dimensional representation. This holds because the two fundamental representations of G_2 are $V_{\omega_1} = std_7$, the irreducible seven-dimensional representation, and $V_{\omega_2} = Lie(G_2)$, the adjoint representation. One knows that

$$\Lambda^2(std_7) = std_7 \oplus Lie(G_2),$$

and hence knowing the characteristic polyomial of an element of UG_2 determines its trace in both $V_{\omega_1} = std_7$ and $V_{\omega_2} = \Lambda^2(std_7) - std_7$, and hence in every irreducible representation of UG_2. By Peter-Weyl, these traces determine the conjugacy class of the element.

Our main work will be to construct, for each finite field k and each element $a \in k^{\times}$, an object $N(a, k) \in \mathcal{P}_{arith}$ on $\mathbb{G}_m/k$ with the following properties.

(1) (extension of scalars): Given $a \in k^{\times}$, and a finite extension field E/k, the pullback of $N(a, k)$ from $\mathbb{G}_m/k$ to $\mathbb{G}_m/E$ is the object $N(a, E)$ on $\mathbb{G}_m/E$ constructed by viewing a as lying in E.

(2) $N(a, k)$ is geometrically Lie-irreducible of "dimension" seven, pure of weight zero, and orthogonally self-dual. It has no bad characters.

(3) For any character χ of $k^{\times}$ (with $\chi \neq \chi_2$ if p is odd), form the lisse sheaf $\mathcal{F}(\chi, k)$ on $\mathbb{G}_m/k$. Then we have the equality of characteristic polynomials

$$\det(1 - TFrob_{k, \chi}|\omega(N(a, k))) = \det(1 - TFrob_{a, k}|\mathcal{F}(\chi, k)).$$

If we grant the existence of such objects $N(a, k)$, we get the following theorem.

Theorem 25.1. *Suppose we have objects $N(a, k)$ as above. Then "with probability one," $N(a, k)$ has $G_{geom} = G_{arith} = G_2$. More precisely, in any sequence of finite fields k_i (possibly of different characteristics) with $\#k_i \to \infty$, the fractions*

$$\frac{\#\{a_i \in k_i^{\times}|N(a_i, k_i) \text{ has } G_{geom} = G_{arith} = G_2\}}{\#k_i^{\times}}$$

tend archimedeanly to 1.

Proof. We will prove this in a series of lemmas.

Lemma 25.2. *Fix a finite field k of characteristic denoted p, and an element $a \in k^\times$. We have the following results concerning the groups G_{geom} and G_{arith} for $N(a,k)$ on $\mathbb{G}_m/k$.*

(1) *We have $G_{geom} = G_{arith} \subset SO(7)$.*

(2) *Either $G_{arith} = G_2$, or G_{arith} is the image of $SL(2)$ in its irreducible representation $V_7 := Sym^6(std_2)$ of dimension seven.*

Proof. Because $N(a,k)$ is orthogonally self-dual, we have $G_{geom} \subset G_{arith} \subset O(7)$. Thus "det"$(N(a,k))$ has order dividing 2, so is either δ_1 or $(-1)^{deg} \otimes \delta_1$, or, if p is odd, possibly δ_{-1} or $(-1)^{deg} \otimes \delta_{-1}$.

We first show that "det"$(N(a,k))$ is δ_1. This will result from the identity

$$\det(1 - TFrob_{E,\rho}|\omega(N(a,k))) = \det(1 - TFrob_{a,E}|\mathcal{F}(\rho, E)),$$

valid for every finite extension E/k and every character ρ of $E^\times$ (except χ_2, if p is odd), together with the fact that each such $Frob_{a,E}|\mathcal{F}(\rho, E)$ lies in $G_2 \subset SO(7)$, so has determinant 1. Thus each such

$$\det(Frob_{E,\rho}|\omega(N(a,k))) = 1.$$

If $p = 2$, and more generally if "det"$(N(a,k))$ is geometrically trivial, take $E = k$ and $\rho = 1$ to eliminate the $(-1)^{deg} \otimes \delta_1$ possibility. If p is odd, then over any sufficiently large extension field E of k ($\#E > 4$ is big enough) there is a character $\rho \neq \chi_2$ with $\rho(-1) = -1$. Taking $\deg(E/k)$ to be of variable parity, the fact that $\det(Frob_{E,\rho}|\omega(N(a,k))) = 1$ eliminates both the δ_{-1} and $(-1)^{deg} \otimes \delta_{-1}$ possibilities. Thus we have $G_{geom} \subset G_{arith} \subset SO(7)$.

Because $N(a,k)$ is geometrically Lie-irreducible, the identity component G^0_{geom} is an irreducible subgroup of $SO(7)$. By Gabber's theorem on prime-dimensional representations [**Ka-ESDE**, 1.6], the only irreducible connected subgroups of $SO(7)$ are $SO(7)$ itself, or G_2 in its seven-dimensional irreducible representation, or the image $SL(2)/\pm 1$ of $SL(2)$ in $V_7 := Sym^6(std_2)$. Each of these groups is its own normalizer in $SO(7)$ (because for each of these groups, every automorphism is inner, and the ambient group $SO(7)$ contains no nontrivial scalars). Therefore G_{arith}, which lies in $SO(7)$ and normalizes G_{geom} and consequently normalizes G^0_{geom}, must be this same group. Thus we have

$$G^0_{geom} = G_{geom} = G_{arith},$$

and G_{arith} is either $SO(7)$ or G_2 or the image $SL(2)/\pm 1$ of $SL(2)$ in $V_7 := Sym^6(std_2)$.

Because $N(a,k)$ in $\mathcal{P}_{arith}$ is geometrically (and hence arithmetically) irreducible, and pure of weight zero, the equality $G_{geom} = G_{arith}$ implies (Theorem 1.1) the equidistribution of the Frobenius conjugacy classes in the space of conjugacy classes of a compact form K of G_{arith}. We now show that G_{arith} is not $SO(7)$. We argue by contradiction. If G_{arith} were $SO(7)$, then K would be $SO(7, \mathbb{R})$ for the Euclidean inner product, its traces would fill the interval $[-5, 7]$, and the set

$$\{g \in K | \mathrm{Tr}(g) < -4\}$$

would be an open set of positive measure. Therefore for large extension fields E/k, there would exist, by equidistribution, characters ρ of $E^\times$ (with $\rho \neq \chi_2$ if p is odd) with $\mathrm{Trace}(Frob_{E,\rho}|\omega(N(a,k))) < -4$. But we have $\mathrm{Trace}(Frob_{E,\rho}|\omega(N(a,k))) = \mathrm{Trace}(Frob_{a,E}|\mathcal{F}(\rho, E))$. As noted above, $Frob_{a,E}|\mathcal{F}(\rho, E)$ has its semisimplification in UG_2, and one knows [**Ka-NotesG2**, 5.5] that the traces of elements of UG_2 in its seven-dimensional irreducible representation lie in the interval $[-2, 7]$.

So the only two possibilities for G_{arith} are G_2 or the image of $SL(2)$ in $V_7 := Sym^6(std_2)$. In the second case, K is the image of $SU(2)$ in $V_7 := Sym^6(std_2)$. □

We next give a property of the elements in the image of $SL(2)$ in $V_7 := Sym^6(std_2)$ which can be used to show that certain elements of G_2 (acting in std_7) do not lie in this image.

Lemma 25.3. *Given an element $g \in GL(7)$, denote by $(u_g, v_g) \in \mathbb{A}^2$ the point*

$$(u_g, v_g) := (\mathrm{Trace}(g), \mathrm{Trace}(g^2)).$$

Denote by $R[U, V] \in \mathbb{Z}[U, V]$ the two-variable polynomial

$$R[U, V] := 8U^3 - 4U^4 - 4U^5 + U^6 + 4U^2V - 3U^4V - 4UV^2 + 3U^2V^2 - V^3.$$

We have the following results.

(1) *For any $\gamma \in SL(2, \mathbb{C})$, with image $g := Sym^6(\gamma) \in SO(V_7)$, we have*

$$R[u_g, v_g] = 0.$$

(2) *There exist elements g in $G_2(\mathbb{C}) \subset SO(7)$ for which*

$$R[u_g, v_g] \neq 0.$$

(3) *In the space $UG_2^{\#}$ of conjugacy classes of UG_2, the set of points $\{g \in UG_2^{\#} | R[u_g, v_g] = 0\}$ is a closed set of (Haar) measure zero.*

Proof. (1) Given $\gamma \in SL(2, \mathbb{C})$, denote by x and $1/x$ its eigenvalues, and by $t := x^2 + 1/x^2$ the trace of its square. The eigenvalues of $g := Sym^6(\gamma)$ are $x^6, x^4, x^2, 1, 1/x^2, 1/x^4, 1/x^6$. In terms of $t = x^2 + 1/x^2$, we have

$$\mathrm{Trace}(g) = 1 + t + (t^2 - 2) + (t^3 - 3t) = -1 - 2t + t^2 + t^3 := f(t).$$

The trace of $(\gamma^2)^2$ is $t^2 - 2$, so we have

$$\text{Trace}(g^2) = f(t^2 - 2) = -1 + 6t^2 - 5t^4 + t^6.$$

The resultant of the two polynomials in t given by

$$f(t) - U, \quad f(t^2 - 2) - V,$$

is, according to Mathematica, $R[U, V]$. Or one verifies by direct substitution that $R[f(t), f(t^2 - 2)] = 0$.

(2) Denote by $\zeta_3 \in \mathbb{C}$ a primitive cube root of unity. There is an element g of G_2 with eigenvalues $1, \zeta_3, \zeta_3, \zeta_3, \overline{\zeta_3}, \overline{\zeta_3}, \overline{\zeta_3}$. For this element, we have $\text{Trace}(g) = \text{Trace}(g^2) = -2$, and one checks that $R[-2, -2] = 216$.

(3) It suffices to show that in the maximal torus UT of UG_2, the locus $R[u_g, v_g] = 0$ has measure zero for Haar measure on UT, since the Herman Weyl measure on UT (i.e., the Weyl group-invariant measure on UT which induces (the direct image of) Haar measure on $UG_2^\# \cong UT/W$) is absolutely continuous with respect to Haar measure. The function $R[u_g, v_g]$ on $UT \cong S^1 \times S^1$ is a trigonometric polynomial, so it is either identically zero or its zero locus has measure zero. By part (2), it is not identically zero. $\qquad\square$

Lemma 25.4. *Let k be a finite field of characteristic denoted p, $a \in k^\times$ an element. Suppose there exists a character χ of $k^\times$, $\chi \neq \chi_2$ if $p \neq 2$, such that the element $\theta_{a,k} := (\text{Frob}_{a,k}|\mathcal{F}(\chi, k))^{ss} \in UG_2^\#$ has $R[u_{\theta_{a,k}}, v_{\theta_{a,k}}] \neq 0$. Then $N(a, k)$ has $G_{geom} = G_{arith} = G_2$.*

Proof. In view of the identity

$$\det(1 - T\text{Frob}_{k,\chi}|\omega(N(a, k))) = \det(1 - T\text{Frob}_{a,k}|\mathcal{F}(\chi, k)),$$

we see that $\theta_{a,k}$ is equal to $(\text{Frob}_{k,\chi}|\omega(N(a, k))^{ss}$. In view of Lemma 25.3, this element in $UG_2^\#$ is not of the form $\text{Sym}^6(\gamma)$ for any element $\gamma \in SU(2)^\#$. This rules out the possibility that $G_{geom} = G_{arith}$ is the image of $SL(2)$ in $\text{Sym}^6(\text{std}_d)$, and the result now follows from Lemma 25.2. $\qquad\square$

Lemma 25.5. *Let k be a finite field, $\mathcal{F}$ a lisse sheaf of rank seven on $\mathbb{G}_m/k$ which is pure of weight zero, orthogonally self-dual, and with $G_{geom} = G_{arith} = G_2$. Denote by $\alpha \in \mathbb{R}_{\geq 0}$ (resp. $\beta \in \mathbb{R}_{\geq 0}$) the largest slope of $\mathcal{F}$ at 0 (resp. ∞). Then for any irreducible nontrivial representation Λ of G_2, we have the estimate*

$$\left| \sum_{a \in k^\times} \text{Trace}(\Lambda(\text{Frob}_{a,k})) \right| \leq (\alpha + \beta) \dim(\Lambda) \sqrt{\#k}.$$

Proof. Consider the lisse sheaf $\Lambda(\mathcal{F})$ of rank $\dim(\Lambda)$ obtained by "pushing out" $\mathcal{F}$ by Λ. By the Lefschetz Trace formula [**Gr-Rat**] and the vanishing of $H_c^2(\mathbb{G}_m/\overline{k}, \Lambda(\mathcal{F}))$, we have

$$\sum_{a \in k^\times} \text{Trace}(\Lambda(Frob_{a,k})) = -\text{Trace}(Frob_k | H_c^1(\mathbb{G}_m/\overline{k}, \Lambda(\mathcal{F}))).$$

By the Euler-Poincaré formula, we have

$$\dim H_c^1(\mathbb{G}_m/\overline{k}, \Lambda(\mathcal{F})) = Swan_0(\Lambda(\mathcal{F})) + Swan_\infty(\Lambda(\mathcal{F})).$$

By assumption, the upper numbering groups $I(0)^{\alpha+}$ and $I(\infty)^{\beta+}$ act trivially on $\mathcal{F}$, so also on $\Lambda(\mathcal{F})$, and so α (resp. β) are upper bounds for the slopes of $\mathcal{F}$ at 0 (resp. ∞). Thus we have

$$Swan_0(\Lambda(\mathcal{F})) \le \alpha \dim(\Lambda), \quad Swan_\infty(\Lambda(\mathcal{F})) \le \beta \dim(\Lambda).$$

By Deligne's main theorem [**De-Weil II**, 3.3.1], applied to the pure of weight zero lisse sheaf $\Lambda(\mathcal{F})$, its H_c^1 is mixed of weight ≤ 1, and the result follows. $\square$

Lemma 25.6. *Let k_i be a sequence of finite fields whose cardinalities q_i tend to ∞. Suppose for each i we have a lisse sheaf $\mathcal{F}_i$ of rank seven on $\mathbb{G}_m/k_i$ which is pure of weight zero, orthogonally self-dual, and with $G_{geom} = G_{arith} = G_2$. Suppose there exist real numbers $\alpha \ge 0$ (resp. $\beta \ge 0$) which are upper bounds for the slopes of every $\mathcal{F}_i$ at 0 (resp. at ∞). Let $Z \subset UG_2^\#$ be a closed set of measure zero (for the induced Haar measure μ) in the space $UG_2^\#$ of conjugacy classes of UG_2. For $a_i \in k_i^\times$, denote by $\theta_{a_i,k_i} \in UG_2^\#$ the Frobenius conjugacy class $(Frob_{a_i,k_i}|\mathcal{F}_i)^{ss}$. Then the fractions*

$$\frac{\#\{a_i \in k_i^\times | \theta_{a_i,k_i} \in Z\}}{\#k_i^\times}$$

tend to 0 as $\#k_i \to \infty$.

Proof. Denote by χ_Z the characteristic function of Z. For each i, denote by μ_i the probability measure on $UG_2^\#$ given by

$$\mu_i := (1/(q_i - 1)) \sum_{a_i \in k_i^\times} \delta_{\theta_{a_i,k_i}}.$$

Then our ratios are the integrals $\int \chi_Z d\mu_i$. Pick a real number $\epsilon > 0$. We will show that for $\#k_i$ sufficiently large, we have $|\int \chi_Z d\mu_i| \le \epsilon$.

   Because Z has μ measure zero, we can find an open set V with $Z \subset V \subset UG_2^\#$ and $\mu(V) \le \epsilon/4$. By Urysohn's Lemma [**Ru**, 2.12], we can find a continuous $\mathbb{R}$-valued function f on $UG_2^{\#}$ with $0 \leq f \leq 1$, with support in V and which is 1 on Z. Notice that

$$0 = \int \chi_Z d\mu \leq \int f d\mu \leq \int \chi_V d\mu \leq \epsilon/4.$$

For each i we have

$$0 \leq \int \chi_Z d\mu_i \leq \int f d\mu_i.$$

So it suffices to show

$$|\int f d\mu_i| \leq \epsilon$$

for $\#k_i$ sufficiently large.

By Peter-Weyl, we can find a finite linear combination of traces of irreducible representations $\Lambda_0 = \mathbb{1}, \Lambda_1, ..., \Lambda_n$ which is uniformly within $\epsilon/4$ of f, say

$$Sup_\theta |f(\theta) - a_0 - \sum_{j=1}^{n} a_j \mathrm{Trace}(\Lambda_j(\theta))| \leq \epsilon/4.$$

Then for every i we have

$$|\int f d\mu_i - \int (a_0 + \sum_{j=1}^{n} a_j \mathrm{Trace}(\Lambda_j(\theta)))d\mu_i| \leq \epsilon/4.$$

This same estimate holds for μ as well. We have seen that $0 \leq \int f d\mu \leq \epsilon/4$, and hence we have

$$|\int (a_0 + \sum_{j=1}^{n} a_j \mathrm{Trace}(\Lambda_j(\theta)))d\mu| \leq \epsilon/2.$$

But in this integral, only the constant term survives, so we get

$$|a_0| \leq \epsilon/2.$$

Now let $A := Sup_j |a_j|$. It follows from the previous lemma, applied separately to each of the finitely many Λ_j, that for $q_i := \#k_i$ sufficiently large,[1] we will have

$$\int \mathrm{Trace}(\Lambda_j(\theta))d\mu_i \leq \epsilon/(4An).$$

Then we get

$$|\int \sum_{j=1}^{n} a_j \mathrm{Trace}(\Lambda_j(\theta))d\mu| \leq \epsilon/4.$$

---

[1] The exact condition is that q_i be large enough that we have the inequalities $(\alpha + \beta)\dim(\Lambda_j)\sqrt{q_i}/(q_i - 1) \leq \epsilon/(4An)$ for $j = 1, ..., n$.

Thus from the above inequality

$$\left| \int f d\mu_i - \int (a_0 + \sum_{j=1}^{n} a_j \text{Trace}(\Lambda_j(\theta))) d\mu_i \right| \leq \epsilon/4$$

we see that

$$\left| \int f d\mu_i \right| \leq \epsilon$$

for $\#k_i$ sufficiently large.   □

Lemma 25.7. *Let k_i be a sequence of finite fields whose cardinalities q_i tend to ∞. For each i take $\mathcal{F}_i := \mathcal{F}(\mathbb{1}, k_i)$ in the previous lemma. Take for Z the closed set $\{g \in UG_2^\# | R[u_g, v_g] = 0\}$. Then the fractions*

$$\frac{\#\{a_i \in k_i^\times | \theta_{a_i, k_i} \in Z\}}{\#k_i^\times}$$

tend to 0 as $\#k_i \to \infty$. Equivalently, the fractions

$$\frac{\#\{a_i \in k_i^\times | R[u_{\theta_{a_i,k_i}}. v_{\theta_{a_i,k_i}}] \neq 0\}}{\#k_i^\times}$$

tend to 1 as $\#k_i \to \infty$.

Proof. Indeed the sheaves $\mathcal{F}_i$ are all tame at 0, and their ∞-slopes are either all $1/7$, in characteristic 2, or they are 0 and six repetitions of $1/6$. So we may take $\alpha = 0$ and $\beta = 1/6$ in the previous lemma.   □

Combining this last lemma with Lemma 25.4, we see that Theorem 25.1 holds.   □

Remark 25.8. It seems plausible that **every** $N(a, k)$ has $G_{geom} = G_{arith} = G_2$. Computer calculations show that this is the case if $k = \mathbb{F}_p$ with $p \leq 100$, for every $a \in \mathbb{F}_p^\times$. Indeed, for these pairs (a, k), the element $\theta_{a,k} := (Frob_{a,k} | \mathcal{F}(\mathbb{1}, k))^{ss} \in UG_2^\#$ has $R[u_{\theta_{a,k}}, v_{\theta_{a,k}}] \neq 0$. [The skeptical reader will correctly object that we are merely avoiding a set of measure zero, namely $R[u_g, v_g] = 0$, hence these calculations are no evidence at all that every $N(a, k)$ has $G_{geom} = G_{arith} = G_2$.]

CHAPTER 26

G_2 Examples: Construction in Characteristic Two

We treat the case of characteristic two separately because it is somewhat simpler than the case of odd characteristic. Recall from the first paragraph of the previous chapter that for k a finite field of characteristic 2, and any character χ of $k^\times$, the Tate-twisted Kloosterman sheaf of rank seven

$$\mathcal{F}(\chi, k) := Kl(\mathbb{1}, \mathbb{1}, \mathbb{1}, \chi, \chi, \overline{\chi}, \overline{\chi})(3)$$

has $G_{geom} = G_{arith} = G_2$. Our first task is to express its stalk at a fixed point $a \in k^\times$ as the finite field Mellin transform of the desired object $N(a, k)$.

We abbreviate

$$Kl_2 := Kl(\mathbb{1}, \mathbb{1}), \quad Kl_3 = Kl(\mathbb{1}, \mathbb{1}, \mathbb{1}).$$

Lemma 26.1. *For a finite extension $k/\mathbb{F}_2$, and for $a \in k^\times$, consider the lisse perverse sheaf on $\mathbb{G}_m \times \mathbb{G}_m/k$, with coordinates (z, t), given by*

$$\mathcal{M} = \mathcal{M}(z, t) := Kl_3(a/(z^2 t)) \otimes Kl_2(z) \otimes Kl_2(zt)(3)[2].$$

Then we have the following results.

(1) *The object $N(a, k) := R(pr_2)_! \mathcal{M}$ on $\mathbb{G}_m/k$ is perverse, of the form $\mathcal{N}_a[1]$ for a sheaf $\mathcal{N}_a$.*

(2) *Denote by $\pi : (\mathbb{G}_m)^3 \to \mathbb{G}_m$ the multiplication map $(x, y, z) \mapsto xyz$. Given a character χ of $k^\times$, consider the lisse perverse sheaf on $(\mathbb{G}_m)^3$ with coordinates x, y, z given by*

$$\mathcal{K}_\chi := Kl_3(x) \otimes Kl_2(y) \otimes \mathcal{L}_{\chi(y)} \otimes Kl_2(z) \otimes \mathcal{L}_{\overline{\chi}(z)}(3)[3].$$

We have an isomorphism of perverse sheaves on $\mathbb{G}_m/k$,

$$\mathcal{F}(\chi, k)[1] \cong R\pi_! \mathcal{K}_\chi.$$

(3) *For any character χ of $k^\times$, we have $H_c^0(\mathbb{G}_m/\overline{k}, N(a, k) \otimes \mathcal{L}_\chi) \cong \mathcal{F}(\chi, k)_a$ as $Frob_k$-module, and $H_c^i(\mathbb{G}_m/\overline{k}, N(a, k) \otimes \mathcal{L}_\chi) = 0$ for $i \neq 0$.*

(4) *The object $N(a, k)$ on $\mathbb{G}_m/k$ is pure of weight zero, lies in $\mathcal{P}_{arith}$, has no bad characters, and is totally wild at both 0 and ∞.*

Proof. (1) The Verdier dual of $\mathcal{M}$ is

$$\overline{\mathcal{M}} := Kl_3(-a/(z^2t)) \otimes Kl_2(z) \otimes Kl_2(zt)(3)[2].$$

But in characteristic 2, this is just $\mathcal{M}$. So the Verdier dual of $N(a, k)$ is $R(pr_2)_\star\mathcal{M}$. Because pr_2 is an affine morphism, and $\mathcal{M}$ is perverse, $R(pr_2)_\star\mathcal{M}$ is semiperverse. So to show that $N(a, k)$ is perverse, it suffices to show that it is semiperverse, i.e., that for all but at most finitely many values of $t_0 \in \overline{k}^\times$, we have

$$H_c^2(\mathbb{G}_m/\overline{k}, Kl_3(a/(z^2t_0)) \otimes Kl_2(z) \otimes Kl_2(zt_0)) = 0.$$

In fact, this H_c^2 vanishes for all t_0, because for each $t_0 \neq 0$, the coefficient sheaf is totally wildly ramified at $z = 0$ (from the $Kl_3(a/(z^2t_0))$ factor, the other two factors being tame at $z = 0$). Thus $N(a, k)[-1]$ is a single sheaf, which we name $\mathcal{N}_a$.

(2) This is simply the expression of the Kloosterman sheaf $\mathcal{F}(\chi, k)$ as an iterated ! multiplicative convolution [**Ka-GKM**, 5.4,5.5], taking into account the isomorphisms $Kl_2(\chi, \chi) \cong Kl_2 \otimes \mathcal{L}_\chi$ and $Kl_2(\overline{\chi}, \overline{\chi}) \cong Kl_2 \otimes \mathcal{L}_{\overline{\chi}}$ [**Ka-ESDE**, 8.2.5].

(3) If we make the substitutions $x = w/(z^2t), y = zt, z = z$, then, using the identity $\chi(zt)\overline{\chi}(z) = \chi(t)$, $\mathcal{K}_\chi$ becomes

$$\mathcal{K}_\chi = Kl_3(w/(z^2t)) \otimes Kl_2(y) \otimes \mathcal{L}_{\chi(t)} \otimes Kl_2(z)(3)[3].$$

The restriction of $\mathcal{K}_\chi[-1]$ to the locus $w = a$, which we view as being $\mathbb{G}_m \times \mathbb{G}_m$ with coordinates z, t, is just $\mathcal{M} \otimes pr_2^\star(\mathcal{L}_\chi)$. So from the isomorphism $\mathcal{F}(\chi, k)[1] \cong R\pi_!\mathcal{K}_\chi$ of part (2), and proper base change, we get

$$\mathcal{F}(\chi, k)_a = R\Gamma_c(\mathbb{G}_m^2/\overline{k}, \mathcal{M} \otimes pr_2^\star(\mathcal{L}_\chi)).$$

This last term is just $R\Gamma_c(\mathbb{G}_m/\overline{k}, N(a, k) \otimes \mathcal{L}_\chi)$, by Leray and the projection formula.

(4) For E/k a finite field extension, and $t_0 \in E^\times$, the stalk of $\mathcal{N}_a$ at t_0 is (by proper base change)

$$H_c^1(\mathbb{G}_m/\overline{k}, Kl_3(a/(z^2t_0)) \otimes Kl_2(z) \otimes Kl_2(zt_0)(3)).$$

As already noted, the coefficient sheaf is totally wildly ramified at $z = 0$.

We first claim that for $t_0 \neq 1$, the coefficient sheaf is also totally wildly ramified at $z = \infty$. As the first factor $Kl_3(a/(z^2t_0))$ is tame, in fact unipotent, at $z = \infty$, we must show that $Kl_2(z) \otimes Kl_2(zt_0)$ is totally wild at $z = \infty$, so long as $t_0 \neq 1$. As $I(\infty)$-representations, $Kl_2(z)$ and its translate $Kl_2(zt_0)$ are both irreducible, so their tensor product is $I(\infty)$-semisimple. In addition, both are $I(\infty)$-self-dual and of trivial determinant. We argue by contradiction. If $Kl_2(z) \otimes Kl_2(zt_0)$ is not totally wild, then it contains some tame character $\mathcal{L}_\chi$ as a summand, and hence we have an $I(\infty)$-isomorphism

$$Kl_2(zt_0) \cong Kl_2(z) \otimes \mathcal{L}_\chi.$$

Taking determinants, we infer that $\chi^2 = 1$. As $p = 2$, this forces $\chi = 1$. But the isomorphism class of $Kl_2(z)$ as $I(\infty)$-representation detects nontrivial translations, hence $t_0 = 1$, contradiction.

For any $t_0 \neq 0$, $Kl_3(a/(z^2 t_0))$ has $Swan_0 = 1$ (because $p = 2$). Hence the coefficient sheaf $Kl_3(a/(z^2 t_0)) \otimes Kl_2(z) \otimes Kl_2(zt_0)$ has $Swan_0 = 4$.

We next show that the Swan conductor at ∞ of $Kl_2(z) \otimes Kl_2(zt_0)$, for $t_0 \neq 1$, is 2. To see this, we argue as follows. Both $Kl_2(z)$ and its translate $Kl_2(zt_0)$ have both ∞-slopes $1/2$, so the four ∞-slopes of $Kl_2(z) \otimes Kl_2(zt_0)$ are each $\leq 1/2$, and each > 0 (by the total wildness). Taking into account the Hasse-Arf theorem, that the multiplicity of any slope is a multiple of its denominator, the only possible slopes are $1/2$ or $1/3$ or $1/4$. No slope can be $1/3$, for then there would be 3 slopes $1/3$, and the remaining slope would be an integer. So either we have all slopes $1/2$, or we have all slopes $1/4$. We argue by contradiction. If we had all slopes $1/4$, then $Kl_2(z) \otimes Kl_2(zt_0)$ would be a lisse sheaf on $\mathbb{G}_m$ which is tame at 0, and totally wild at ∞ with $Swan_\infty = 1$. Any such sheaf is a multiplicative translate of a Kloosterman sheaf, cf. [**Ka-GKM**, 8.7.1]. The local monodromy at 0 is $Unip(2) \otimes Unip(2) \cong Unip(3) \oplus Unip(1)$, but the only Kloosterman sheaves which are unipotent at 0 have their local monodromy at 0 a single Jordan block.

Hence the coefficient sheaf $Kl_3(a/(z^2 t_0)) \otimes Kl_2(z) \otimes Kl_2(zt_0)$ has $Swan_\infty = 6$, for $t_0 \neq 1$. Thus $\mathcal{N}_a|\mathbb{G}_m \setminus \{1\}$ is lisse of rank 10 (by Deligne's semicontinuity theorem [**Lau-SCCS**, 2.1.2]) and pure of weight -1.

We next claim that $\mathcal{N}_a$ on $\mathbb{G}_m$ is a middle extension, i.e., that for $j_1 : \mathbb{G}_m \setminus \{1\} \subset \mathbb{G}_m$ the inclusion, we have $\mathcal{N}_a \cong j_{1\star} j_1^\star \mathcal{N}_a$. To see this, recall that $N(a,k) = \mathcal{N}_a[1]$ is perverse, so the adjunction map is injective: $\mathcal{N}_a \subset j_{1\star} j_1^\star \mathcal{N}_a$ as sheaves. So we have a short exact sequence of sheaves with a punctual third term supported at 1, say

$$0 \to \mathcal{N}_a \to j_{1\star} j_1^\star \mathcal{N}_a \to \delta_1 \otimes V \to 0,$$

with V a $Gal(\overline{k}/k)$-module. Because $\mathcal{N}_a|\mathbb{G}_m \setminus \{1\}$ is lisse and pure of weight -1 (in fact, we only "need" that it is lisse and mixed of weight ≤ -1), V is mixed of weight ≤ -1, cf. [**De-Weil II**, 1.8.1]. In the long exact cohomology sequence with compact supports, the group $H_c^0(\mathbb{G}_m/\overline{k}, j_{1\star} j_1^\star \mathcal{N}_a) = 0$ (as the sheaf $j_{1\star} j_1^\star \mathcal{N}_a$ has no nonzero punctual sections), so we get an injection $V \subset H_c^1(\mathbb{G}_m/\overline{k}, \mathcal{N}_a)$. By part (3), this last group is the stalk $\mathcal{F}(\mathbb{1}, k)_a$, which is pure of weight 0. Therefore $V = 0$.

Therefore $N(a, k) = \mathcal{N}_a[1]$ is the middle extension from $\mathbb{G}_m \setminus \{1\}$ of a (lisse) perverse sheaf which is pure of weight 0, hence is itself pure of weight 0. The isomorphism of part (3), applied also to all extensions of scalars $N(a, E)$, shows that for any finite extension field E/k, and for any character χ of $E^\times$, the group $H^1_c(\mathbb{G}_m/\overline{k}, \mathcal{N}_a \otimes \mathcal{L}_\chi) \cong \mathcal{F}(\chi, E)_a$ is pure of weight 0, and the H^2_c vanishes. Therefore $N(a, k)$ has, geometrically, no quotient which is a Kummer sheaf $\mathcal{L}_\chi[1]$. But as $N(a, k)$ is pure, it is geometrically semisimple, so it has, geometrically, no Kummer subsheaf either. Hence $N(a, k)$ lies in $\mathcal{P}_{arith}$. By Theorem 4.1, the purity of $H^1_c(\mathbb{G}_m/\overline{k}, \mathcal{N}_a \otimes \mathcal{L}_\chi)$ for every χ implies that $N(a, k)$ has no bad characters, which, as already noted, is equivalent to the fact that $\mathcal{N}_a$ is totally wild at both 0 and ∞. $\qquad\square$

It remains to show that the object $N(a, k)$ we have constructed in characteristic 2 is orthogonally self-dual, of "dimension" seven, and geometrically Lie-irreducible. That it is of "dimension" seven is obvious from part (3) of the previous lemma, applied with any one choice of χ. We next show that $N(a, k)$ is self-dual as an object of $\mathcal{P}_{arith}$.

Lemma 26.2. *We have the following results.*

(1) *The object $N(a, k)$ is isomorphic to its pullback by multiplicative inversion $t \mapsto 1/t$ on $\mathbb{G}_m$.*

(2) *The object $N(a, k)$ is isomorphic to its Verdier dual $D(N(a, k))$.*

(3) *As an object in the Tannakian category $\mathcal{P}_{arith}$, $N(a, k)$ is self-dual.*

Proof. Assertion (3) is immediate from (1) and (2), since $N(a, k)^\vee \cong [t \mapsto 1/t]^\star D(N(a, k))$.

Assertion (1) is obvious from the definition of $N(a, k)$ given in part (1) of the previous lemma, by the change of variable $z \mapsto z/t, t \mapsto t$ in the sheaf $\mathcal{M} := Kl_3(a/(z^2 t)) \otimes Kl_2(z) \otimes Kl_2(zt)(3)[2]$ on $\mathbb{G}_m \times \mathbb{G}_m$. It remains to show (2), that $N(a, k)$ is its own Verdier dual. It suffices to show that this holds over the open set $\mathbb{G}_m \setminus \{1\}$, since middle extension of perverse sheaves from $\mathbb{G}_m \setminus \{1\}$ to $\mathbb{G}_m$ commutes with Verdier duality. On $\mathbb{G}_m \times \mathbb{G}_m$, the coefficient sheaf $\mathcal{M}$ is self-dual (remember we are in characteristic 2, so $Kl_3(1)$ is self-dual), so the Verdier dual of $N(a, k) := R(pr_2)_! \mathcal{M}$ is $R(pr_2)_\star \mathcal{M}$. It suffices to show that the canonical "forget supports" map

$$R(pr_2)_! \mathcal{M} \to R(pr_2)_\star \mathcal{M}$$

is an isomorphism over $\mathbb{G}_m \setminus \{1\}$. As we have seen in the proof of the previous lemma, $R(pr_2)_! \mathcal{M}$ is lisse over $\mathbb{G}_m \setminus \{1\}$. Therefore its

Verdier dual, $R(pr_2)_\star \mathcal{M}$, is also lisse over $\mathbb{G}_m \setminus \{1\}$, and its formation is compatible with arbitrary change of base on $\mathbb{G}_m \setminus \{1\}$. So to show that the "forget supports" map is an isomorphism over $\mathbb{G}_m \setminus \{1\}$, it suffices to check point by point in $\mathbb{G}_m \setminus \{1\}$. We have already observed that over any point $t_0 \in \mathbb{G}_m \setminus \{1\}$, the restriction of $\mathcal{M}(-3)[-2]$ to that fibre, namely the lisse sheaf $Kl_3(a/(z^2 t_0)) \otimes Kl_2(z) \otimes Kl_2(z t_0)$, is totally wild at both 0 and ∞, and hence we have the desired isomorphism $H_c \cong H$ of its cohomology groups. $\qquad \square$

Although the "dimension" of $N(a, k)$ is odd, namely seven, we do not yet know that $N(a, k)$ is irreducible, much less that it is geometrically Lie-irreducible. So we cannot assert that the autoduality we have shown $N(a, k)$ to admit is necessarily orthogonal (or indeed that it has a sign at all). To clarify this question, we need information on the local monodromy of the sheaf $\mathcal{N}_a = N(a, k)[-1]$ at the point 1.

Lemma 26.3. *We have the following results.*

(1) *The stalk*

$$(\mathcal{N}_a(-3))_1 = H_c^1(\mathbb{G}_m/\overline{k}, Kl_3(a/z^2) \otimes Kl_2(z) \otimes Kl_2(z))$$

has rank seven. It has six Frobenius eigenvalues of weight 5, and one Frobenius eigenvalue of weight 2.

(2) *As $I(1)$-representation, $\mathcal{N}_a$ is $Unip(1)^6 \oplus Unip(4)$.*

(3) *The sheaf $\mathcal{N}_a$ is totally wild at both 0 and ∞, with $Swan_0 = Swan_\infty = 2$.*

Proof. (1) Because we are in characteristic 2, we have the Carlitz isomorphism

$$Kl_2(z) \otimes Kl_2(z) \cong \overline{\mathbb{Q}_\ell}(-1) \oplus Kl_3(z),$$

cf. [**Ka-ClausCar**, 3.1]. So our coefficient sheaf is

$$Kl_3(a/z^2)(-1) \bigoplus Kl_3(a/z^2) \otimes Kl_3(z).$$

Because we are in characteristic two, the map $z \mapsto a/z^2$ is radicial and bijective, so we have

$$H_c^1(\mathbb{G}_m/\overline{k}, Kl_3(a/z^2)(-1)) \cong H_c^1(\mathbb{G}_m/\overline{k}, Kl_3(z)(-1)) = \overline{\mathbb{Q}_\ell}(-1),$$

the last equality by Mellin inversion, cf. [**Ka-GKM**, 4.0]. This is the eigenvalue of weight 2. The sheaf $Kl_3(a/z^2) \otimes Kl_3(z)$ is lisse on $\mathbb{G}_m$, pure of weight 4, totally wild at both 0 and ∞ with all slopes $1/3$ (remember we are in characteristic 2, so $Kl_3(a/z^2)$ still has its $I(0)$-representation totally wild with Swan conductor 1, rather than the 2 it would be in odd characteristic). So $Swan_0 = Swan_\infty = 3$, and the cohomology group $H^1_c(\mathbb{G}_m/\overline{k}, Kl_3(a/z^2) \otimes KL_3(z))$ has dimension six and is pure of weight 5.

(2) Since $\mathcal{N}_a(-3)$ is the middle extension across 1 of a lisse sheaf on $\mathbb{G}_m \setminus \{1\}$ which is pure of weight 5, the weights of the Frobenius eigenvalues on the space of $I(1)$-invariants tell us that the unipotent part of the $I(1)$-representation is precisely $Unip(1)^6 \oplus Unip(4)$, cf. [**De-Weil II**, 1.8.4 and 1.7.14.2-3] or [**Ka-GKM**, 7.0.7]. As this lisse sheaf $\mathcal{N}_a|\mathbb{G}_m \setminus \{1\}$ has rank 10, there is room for nothing more in its $I(1)$-representation.

(3) From the fact that $N(a, k)$ has no bad chararcters, we know that $\mathcal{N}_a$ is totally wild at both 0 and ∞. From its invariance under $t \mapsto 1/t$, we know that $Swan_0 = Swan_\infty$. Because $\mathcal{N}_a$ is lisse outside 1 and unipotent (hence tame) there, we have the equation

$$Swan_0(\mathcal{N}_a) + Swan_\infty(\mathcal{N}_a) + drop_1(\mathcal{N}_a) = \text{``}dim\text{''}(N(a, k)) = 7.$$

But $drop_1 = 3$, so $Swan_0 + Swan_\infty = 4$. As $Swan_0 = Swan_\infty$, we are done.   □

Lemma 26.4. *The object $N(a, k)$ is geometrically Lie-irreducible, and orthogonally self-dual.*

Proof. It suffices to prove the first statement, since $N(a, k)$ is self-dual and of "dimension" seven. It suffices to show that $N(a, k)$ is geometrically irreducible, since it has a unique singularity, namely 1, in $\mathbb{G}_m$, so cannot be geometrically isomorphic to any nontrivial multiplicative translate of itself.

We argue by contradiction. Suppose we have, geometrically, a direct sum decomposition

$$N(a, k) \cong \mathcal{A}[1] \oplus \mathcal{B}[1].$$

Then both $\mathcal{A}$ and $\mathcal{B}$ must be totally wild at both 0 and ∞, so each of the four Swan conductors (of $\mathcal{A}$ and $\mathcal{B}$ at 0 and ∞) is ≥ 1. As their direct sum $\mathcal{A} \oplus \mathcal{B} = \mathcal{N}_a$ has $Swan_0 = Swan_\infty = 2$, we must have

$$Swan_0(\mathcal{A}) = Swan_\infty(\mathcal{A}) = 1, \quad Swan_0(\mathcal{B}) = Swan_\infty(\mathcal{B}) = 1.$$

Furthermore, both $\mathcal{A}$ and $\mathcal{B}$ are lisse on $\mathbb{G}_m \setminus \{1\}$, and middle extensions across 1. So both $\mathcal{A}[1]$ an $\mathcal{B}[1]$ are geometrically irreducible. Their local monodromies at 0 are both totally wild of Swan conductor 1, so detect nontrivial multiplicative translations. So each of $\mathcal{A}[1]$ and $\mathcal{B}[1]$ is geometrically Lie-irreducible. The local monodromy at 1 of $\mathcal{N}_a$ is $Unip(1)^6 \oplus Unip(4)$, so one of our summands is lisse at 1, say $\mathcal{A}$ at 1 is $Unip(1)^a$, and the other is $\mathcal{B} = Unip(1)^b \oplus Unip(4)$. Thus $\mathcal{A}[1]$ has "dimension" 2, and $\mathcal{B}[1]$ has "dimension" 5. As they have different "dimensions" and are geometrically irreducible, any autoduality of

$N(a, k)$ must make each of the summands self-dual. For $\mathcal{B}[1]$, the auto-duality must be orthogonal, since its "dimension" is odd. For $\mathcal{A}[1]$, the autoduality must be symplectic, for otherwise its G_{geom} lies in $O(2)$, which contradicts its Lie-irreducibility.

Since G_{geom} is a normal subgroup of G_{arith}, G_{arith} must permute the G_{geom}-irreducible constituents of $\omega(N(a, k))$. As these two con-stituents have different dimensions, each must be G_{arith}-stable. Thus $N(a, k) = \mathcal{A}[1] \oplus \mathcal{B}[1]$ is an arithmetic direct sum decomposition into irreducibles of "dimensions" 2 and 5. As $N(a, k)$ admits an arithmetic autoduality, this autoduality must make each summand self-dual, in a way compatible with its geometric autoduality. Thus $\mathcal{A}[1]$ is sym-plectically self-dual, and $\mathcal{B}[1]$ is orthogonally self-dual. So $G_{arith,\mathcal{A}[1]} \subset SL(2)$, and $G_{arith,\mathcal{B}[1]} \subset O(5)$. In fact, $G_{arith,\mathcal{B}[1]} \subset SO(5)$, because "det"$(N(a, k))$ is trivial. Indeed, by Lemma 26.1, part (3), every Frobenius $Frob_{E,\chi}$ of $N(a, k)$ lies inside a G_2 inside an $SO(7)$. But "det"$(N(a, k)) \cong$ "det"$(\mathcal{A}[1]) \star_{mid}$ "det"$(\mathcal{B}[1]) \cong$ "det"$(\mathcal{B}[1])$ (this last because "det"$(\mathcal{A}[1])$ is trivial, as $G_{arith,\mathcal{A}[1]} \subset SL(2)$).

As $\mathcal{A}[1]$ is geometrically Lie-irreducible with its $G_{arith,\mathcal{A}[1]} \subset SL(2)$, $G^0_{geom,\mathcal{A}[1]}$ is an irreducible connected subgroup of $SL(2)$, so must be $SL(2)$, so we have

$$G_{geom,\mathcal{A}[1]} = G_{arith,\mathcal{A}[1]} = SL(2).$$

As $\mathcal{B}[1]$ is geometrically Lie-irreducible with its $G_{arith,\mathcal{B}[1]} \subset SO(5)$, $G^0_{geom,\mathcal{B}[1]}$ is an irreducible connected subgroup of $SO(5)$. By Gab-ber's theorem on prime-dimensional representations, either $G^0_{geom,\mathcal{B}[1]}$ is $SO(5)$ or it is the image $SL(2)/\pm 1$ of $SL(2)$ in $SO(5)$ via $Sym^4(std_2)$. Both of these groups are their own normalizers in the ambient $SO(5)$, so

$$G_{geom,\mathcal{B}[1]} = G_{arith,\mathcal{B}[1]} = SO(5) \text{ or } G_{geom,\mathcal{B}[1]} = G_{arith,\mathcal{B}[1]} = SL(2)/\pm 1.$$

Now we apply Goursat's lemma to conclude that we have one of three possibilities. Either

$$G_{geom,N(a,k)} = G_{arith,N(a,k)} = SL(2) \times SO(5),$$

or

$$G_{geom,N(a,k)} = G_{arith,N(a,k)} = SL(2) \times SL(2)/\pm 1,$$

or

$$G_{geom,N(a,k)} = G_{arith,N(a,k)} = SL(2).$$

In this last case, $SL(2)$ acts by the representation $std_2 \oplus Sym^4(std_2)$, i.e., this $SL(2)$ is the subgroup of $SL(2) \times SL(2)/\pm 1$ which is the graph of the projection of the first factor onto the second.

Even for this smallest group, its compact form contains elements $Diag(e^{i\theta}, e^{-i\theta})$ with, e.g., $\theta = 3\pi/4$, whose trace in $std_2 \oplus Sym^4(std_2)$, namely

$$2\cos(\theta) + 1 + 2\cos(2\theta) + 2\cos(4\theta),$$

is $-1 - \sqrt{2} = -2.414... < -2.4$. So in the compact form K of any of the three possible groups $G_{arith,N(a,k)}$, the set "Trace < -2.4" is a nonempty open set in the compact space $K^{\#}$. So by equidistribution we will find Frobenius elements $Frob_{E,\chi}$ for $N(a,k)$ whose traces are < -2.4. But each such Frobenius is, by Lemma 26.1, part (3), an element of UG_2, and so its trace lies in $[-2, 7]$. This contradiction completes the proof that $N(a, k)$ is in fact geometrically irreducible. □

This concludes the construction of the objects $N(a, k)$ in characteristic 2.

CHAPTER 27

G_2 Examples: Construction in Odd Characteristic

The situation in odd characteristic is complicated by the quadratic character χ_2: the ! hypergeometric sheaf of type $(7, 1)$

$$\mathcal{F}(\chi_2, k) := (A^{-7})^{deg} \otimes \mathcal{H}(\mathbb{1}, \mathbb{1}, \mathbb{1}, \chi_2, \chi_2, \chi_2, \chi_2; \chi_2),$$

though lisse on $\mathbb{G}_m$, is **not** pure of weight zero, **nor** is its G_{geom} the group G_2. One knows [**Ka-ESDE**, 8.4.7] that this sheaf sits in a short exact sequence

$$0 \to V \otimes \mathcal{L}_{\chi_2} \to \mathcal{F}(\chi_2, k) \to (A^{-7})^{deg} \otimes Kl(\mathbb{1}, \mathbb{1}, \mathbb{1}, \chi_2, \chi_2, \chi_2)(-1) \to 0,$$

with

$$V := (A^{-7})^{deg} \otimes H^1_c(\mathbb{G}_m/\overline{k}, Kl(\mathbb{1}, \mathbb{1}, \mathbb{1}, \chi_2, \chi_2, \chi_2)).$$

The rank 6 quotient $(A^{-7})^{deg} \otimes Kl(\mathbb{1}, \mathbb{1}, \mathbb{1}, \chi_2, \chi_2, \chi_2)(-1)$ is pure of weight 0, and the rank one subobject $V \otimes \mathcal{L}_{\chi_2}$ is pure of weight -4; in fact $Frob_k|V = A^{-4}$, cf. [**Ka-GKM**, 4.0].

In order to construct the objects $N(a, k)$, in odd characteristic we will first construct an "approximation" $N_0(a, k)$ to it, then "refine" this approximation. We abbreviate

$$Kl_2 = Kl(\mathbb{1}, \mathbb{1}), \quad \mathcal{H}_{3,1} = \mathcal{H}(\mathbb{1}, \mathbb{1}, \mathbb{1}; \chi_2).$$

We have the following version of Lemma 26.1.

Lemma 27.1. *For a finite extension $k/\mathbb{F}_p$, p odd, and for $a \in k^\times$, consider the lisse perverse sheaf on $\mathbb{G}_m \times \mathbb{G}_m/k$, with coordinates (z, t), given by*

$$\mathcal{M} = \mathcal{M}(z, t) := (A^{-7})^{deg} \otimes \mathcal{H}_{3,1}(a/(z^2 t)) \otimes Kl_2(z) \otimes Kl_2(zt)[2].$$

Then we have the following results.

(1) *The object $N_0(a, k) := R(pr_2)_! \mathcal{M}$ on $\mathbb{G}_m/k$ is perverse, of the form $\mathcal{G}_a[1]$ for a sheaf $\mathcal{G}_a$.*

(2) *Denote by $\pi : (\mathbb{G}_m)^3 \to \mathbb{G}_m$ the multiplication map $(x, y, z) \mapsto xyz$. Given a character χ of $k^\times$, consider the lisse perverse sheaf on $(\mathbb{G}_m)^3$ with coordinates x, y, z given by*

$$\mathcal{K}_\chi := (A^{-7})^{deg} \otimes \mathcal{H}_{3,1}(x) \otimes Kl_2(y) \otimes \mathcal{L}_{\chi(y)} \otimes Kl_2(z) \otimes \mathcal{L}_{\overline{\chi}(z)}[3].$$

We have an isomorphism of perverse sheaves on $\mathbb{G}_m/k$,

$$\mathcal{F}(\chi, k)[1] \cong R\pi_! \mathcal{K}_\chi.$$

(3) *For any character χ of $k^\times$, we have $H^0_c(\mathbb{G}_m/\overline{k}, N_0(a, k) \otimes \mathcal{L}_\chi) \cong$
$\mathcal{F}(\chi, k)_a$ as $Frob_k$-module, and $H^i_c(\mathbb{G}_m/\overline{k}, N_0(a, k) \otimes \mathcal{L}_\chi) = 0$
for $i \neq 0$.*

Proof. (1) The object $\mathcal{M}$ is its own Verdier dual, so exactly as in the
proof of Lemma 26.1 it suffices to show that $R(pr_2)_!\mathcal{M}$ is semiperverse,
i.e., that for all but at most finitely many values of $t_0 \in \overline{k}^\times$, we have

$$H^2_c(\mathbb{G}_m/\overline{k}, \mathcal{H}_{3,1}(a/(z^2 t_0)) \otimes Kl_2(z) \otimes Kl_2(zt_0)) = 0.$$

In fact, this H^2_c vanishes for all t_0. For $t_0 \neq 1$, the H^2_c vanishes because
the coefficient sheaf is totally wild at ∞ with all slopes $1/2$, from the
$Kl_2(z) \otimes Kl_2(zt_0)$ factor, the $\mathcal{H}_{3,1}$ factor being tame $(Unip(3))$ at ∞.
At $t_0 = 1$, the coefficient sheaf is

$$\mathcal{H}_{3,1}(a/z^2) \otimes Kl_2(z) \otimes Kl_2(z).$$

For the last two factors, we have [**Ka-ClausCar**, 3.5]

$$Kl_2(z) \otimes Kl_2(z)(-1) \cong \overline{\mathbb{Q}}_\ell(-2) \bigoplus A^{deg} \otimes \mathcal{H}_{3,1}(4z).$$

So we must show the vanishing of the H^2_c with coefficients in $\mathcal{H}_{3,1}(a/z^2)$
and in $\mathcal{H}_{3,1}(a/z^2) \otimes \mathcal{H}_{3,1}(4z)$. The first vanishes because $\mathcal{H}_{3,1}(a/z^2)$ is
geometrically irreducible of rank 3. The second vanishes because both
$\mathcal{H}_{3,1}(a/z^2)$ and $\mathcal{H}_{3,1}(4z)$ are geometrically irreducible and geometrically
self-dual, but they are not isomorphic to each other : e.g., $\mathcal{H}_{3,1}(a/z^2)$ is
tame at ∞, while $\mathcal{H}_{3,1}(4z)$ is not (being the direct sum, geometrically,
of $\mathcal{L}_{\chi_2}$ and a wild part of rank 2 with both slopes $1/2$, cf. [**Ka-ESDE**,
8.4.11]). Thus $N_0(a, k)[-1]$ is a single sheaf, which we name $\mathcal{G}_a$.

(2) Exactly as in the proof of Lemma 26.1, this is the expression
of the hypergeometric sheaf $\mathcal{F}(\chi, k)$ as the iterated ! multiplicative
convolution (up to a shift and a twist) of $Kl_2(\chi, \chi) \cong Kl_2 \otimes \mathcal{L}_\chi$,
$Kl_2(\overline{\chi}, \overline{\chi}) \cong Kl_2 \otimes \mathcal{L}_{\overline{\chi}}$, and $\mathcal{H}_{3,1}$, cf. [**Ka-ESDE**, 8.2].

(3) The proof is identical to that of part (3) of Lemma 26.1 if we
replace Kl_3 there by $\mathcal{H}_{3,1}$, and replace the Tate twist (3) there by the
$(A^{-7})^{deg}$ twist. $\qquad\square$

Lemma 27.2. *We have the following results concerning the perverse
sheaf $N_0(a, k) = \mathcal{G}_a[1]$.*

(1) *The restriction to $\mathbb{G}_m \setminus \{1\}$ of $N_0(a, k)$ is lisse of rank 14, and
mixed of weight ≤ 0. Its associated graded pieces for the weight
filtration have ranks*

$$rk(gr^0_W N_0(a, k)) = 12, \ \ rk(gr^{-1}_W N_0(a, k)) = 1, rk(gr^{-3}_W N_0(a, k)) = 1.$$

(2) *The restriction to* $\mathbb{G}_m \setminus \{1\}$ *of* $\mathcal{G}_a$ *is lisse of rank* 14, *and mixed of weight* ≤ -1. *Its associated graded pieces for the weight filtration have ranks*

$$rk(gr_W^{-1}\mathcal{G}_a) = 12, \ rk(gr_W^{-2}\mathcal{G}_a) = 1, rk(gr_W^{-4}\mathcal{G}_a) = 1.$$

(3) *For* $j_1 : \mathbb{G}_m \setminus \{1\} \subset \mathbb{G}_m$ *the inclusion, the adjunction map is an isomorphism*

$$\mathcal{G}_a \cong j_{1\star}j_1^\star\mathcal{G}_a.$$

Proof. (1) and (2) are equivalent. For $t_0 \neq 1$, the stalk at t_0 of $(A^{-7})^{deg} \otimes \mathcal{G}_a$ is the group

$$H_c^1(\mathbb{G}_m/\overline{k}, \mathcal{H}_{3,1}(a/(z^2t_0)) \otimes Kl_2(z) \otimes Kl_2(zt_0)) = 0.$$

As already noted in the proof of the previous lemma, the coefficient sheaf is totally wild at ∞, with all slopes $1/2$ (from the $Kl_2(z) \otimes Kl_2(zt_0)$ factor), so with $Swan_\infty = 6$. At 0, the local monodromy is

$$(Unip(1) \oplus (\text{tot. wild, rk} = 2, \text{ slopes both } 1)) \otimes Unip(2) \otimes Unip(2).$$

We have $Unip(2) \otimes Unip(2) \cong Unip(1) \oplus Unip(3)$, so the local monodromy at 0 is

$$= Unip(1) \oplus Unip(3) \oplus (\text{tot. wild, rk} = 8, \text{ slopes all } 1).$$

Thus $Swan_0 = 8$. So each stalk of $\mathcal{G}_a|\mathbb{G}_m \setminus \{1\}$ has rank 14, and its lisseness there results from Deligne's semicontinuity theorem [**Lau-SCCS**, 2.1.2].

The weight filtration of a lisse mixed sheaf on a smooth scheme, here $\mathbb{G}_m \setminus \{1\}$, is a filtration by lisse subsheaves, so we can read the ranks of the associated graded pieces at any chosen point $t_0 \neq 1$. Over such a point, the coefficient sheaf $(A^{-7})^{deg} \otimes \mathcal{H}_{3,1}(a/(z^2t_0)) \otimes Kl_2(z) \otimes Kl_2(zt_0)$ on $\mathbb{G}_m$ is lisse, pure of weight -2, totally wild at ∞, and with the unipotent part of its local monodromy at 0 given by $Unip(1) \oplus Unip(3)$. So we have one weight drop of 1 and one weight drop of 3, from the "expected" weight of -1, cf. Theorem 16.1.

(3) The proof is identical to the proof of this same statement in characteristic 2, where it was proved in the course of proving part (4) of Lemma 26.1. □

We can now define the desired object $N(a, k)$ in the odd characteristic case:

$$N(a, k) := gr_W^0 N_0(a, k).$$

Thus we have a short exact sequence of perverse sheaves on $\mathbb{G}_m$

$$0 \to gr_W^{<0} N_0(a, k) \to N_0(a, k) \to N(a, k) \to 0.$$

Lemma 27.3. *The perverse sheaf $gr_W^{\leq 0} N_0(a, k)$ is negligible, i.e., its Euler characteristic on $\mathbb{G}_m/\overline{k}$ vanishes.*

Proof. For any perverse sheaf M on $\mathbb{G}_m/\overline{k}$, there are only finitely many Kummer sheaves $\mathcal{L}_\chi$ for which $H_c^1(\mathbb{G}_m/\overline{k}, M \otimes \mathcal{L}_\chi)$ is nonzero. One sees this by reduction to the case when M is irreducible. Then unless M is an $\mathcal{L}_\chi[1]$ itself, there are no such. So for general χ, we have a short exact sequence

$$0 \to H_c^0(\mathbb{G}_m/\overline{k}, gr_W^{\leq 0} N_0(a, k) \otimes \mathcal{L}_\chi) \to$$
$$\to H_c^0(\mathbb{G}_m/\overline{k}, N_0(a, k) \otimes \mathcal{L}_\chi) \to H_c^0(\mathbb{G}_m/\overline{k}, N(a, k) \otimes \mathcal{L}_\chi) \to 0.$$

The middle H_c^0 is pure of weight 0 for any $\chi \neq \chi_2$, whereas the first H_c^0 is mixed of lower weight, so it must vanish. But for general χ all the H_c^1 vanish, and in any case only H_c^0 and H_c^1 are possibly nonzero, cf. Lemma 3.5. Hence for general χ the object $gr_W^{\leq 0} N_0(a, k) \otimes \mathcal{L}_\chi$ has vanishing cohomology on $\mathbb{G}_m/\overline{k}$, so in particular vanishing Euler characteristic. But Euler characteristic on $\mathbb{G}_m/\overline{k}$ is invariant under tensoring with a Kummer sheaf, so $gr_W^{\leq 0} N_0(a, k)$ itself has vanishing Euler characteristic. $\qquad\square$

Lemma 27.4. *For any finite extension field E/k, and any character χ of $E^\times$ with $\chi \neq \chi_2$, we have the following results.*

(1) *The groups $H_c^i(\mathbb{G}_m/\overline{k}, gr_W^{\leq 0} N_0(a, k) \otimes \mathcal{L}_\chi)$ vanish for all i.*

(2) *The group $H_c^0(\mathbb{G}_m/\overline{k}, N(a, k) \otimes \mathcal{L}_\chi)$ has rank 7 and is pure of weight zero, and the other $H_c^i(\mathbb{G}_m/\overline{k}, N(a, k) \otimes \mathcal{L}_\chi)$ vanish.*

(3) *The canonical map*

$$H_c^0(\mathbb{G}_m/\overline{k}, N_0(a, k) \otimes \mathcal{L}_\chi)) \to H_c^0(\mathbb{G}_m/\overline{k}, N(a, k) \otimes \mathcal{L}_\chi)$$

is an isomorphism.

(4) *There is a $Frob_E$-isomorphism*

$$H_c^0(\mathbb{G}_m/\overline{k}, N(a, k) \otimes \mathcal{L}_\chi) \cong \mathcal{F}(\chi, E)_a.$$

Proof. In view of Lemma 3.5, we have, for **any** χ, including χ_2, a six term exact sequence

$$0 \to H_c^0(\mathbb{G}_m/\overline{k}, gr_W^{\leq 0} N_0(a, k) \otimes \mathcal{L}_\chi) \to H_c^0(\mathbb{G}_m/\overline{k}, N_0(a, k) \otimes \mathcal{L}_\chi)$$
$$\to H_c^0(\mathbb{G}_m/\overline{k}, N(a, k) \otimes \mathcal{L}_\chi) \to H_c^1(\mathbb{G}_m/\overline{k}, gr_W^{\leq 0} N_0(a, k) \otimes \mathcal{L}_\chi)$$
$$\to H_c^1(\mathbb{G}_m/\overline{k}, N_0(a, k) \otimes \mathcal{L}_\chi) \to H_c^1(\mathbb{G}_m/\overline{k}, N(a, k) \otimes \mathcal{L}_\chi) \to 0.$$

By part (3) of Lemma 27.1, the fifth term vanishes, hence also the sixth term, which (together with Lemma 3.5) gives the vanishing asserted in part (2) for any χ. For $\chi \neq \chi_2$, the second term is pure of weight 0, while the first term is mixed of lower weight, so must vanish. As

$gr_W^{\leq 0} N_0(a, k)$ is negligible, the fourth term consequently vanishes as well. Parts (1), (2), and (3) are now obvious, and we get (4) from (3) and from part (4) of Lemma 27.1. $\qquad\square$

Lemma 27.5. *The object $N(a, k)$ lies in $\mathcal{P}_{arith}$, and has "dimension" 7. It has no bad characters.*

Proof. By construction, $N(a, k)$ is pure of weight 0, hence geometrically semisimple. So if it were not in $\mathcal{P}$, it would have, geometrically, a direct summand $\mathcal{L}_\chi[1]$, and hence would have a nonzero $H_c^1(\mathbb{G}_m/\overline{k}, N(a, k) \otimes \mathcal{L}_{\overline{\chi}})$. But as noted in the proof of the previous lemma, this H_c^1 vanishes for every χ. That $N(a, k)$ has "dimension" 7 follows from part (2) of the previous lemma.

It remains to prove that $N(a, k)$ has no bad characters. In view of the previous lemma, the only possibly bad character is χ_2. The six term exact sequence of the previous lemma becomes a four term exact sequence

$$0 \to H_c^0(\mathbb{G}_m/\overline{k}, gr_W^{\leq 0} N_0(a, k) \otimes \mathcal{L}_{\chi_2}) \to H_c^0(\mathbb{G}_m/\overline{k}, N_0(a, k) \otimes \mathcal{L}_{\chi_2})$$
$$\to H_c^0(\mathbb{G}_m/\overline{k}, N(a, k) \otimes \mathcal{L}_{\chi_2}) \to H_c^1(\mathbb{G}_m/\overline{k}, gr_W^{\leq 0} N_0(a, k) \otimes \mathcal{L}_{\chi_2}) \to 0.$$

By part (3) of Lemma 27.1 and the analysis of the sheaf $\mathcal{F}(\chi_2, k)$ given at the beginning of this chapter, the group $H_c^0(\mathbb{G}_m/\overline{k}, N_0(a, k) \otimes \mathcal{L}_{\chi_2})$ has six eigenvalues of weight 0 and one eigenvalue of weight -4.

We first show that the group $H_c^0(\mathbb{G}_m/\overline{k}, gr_W^{\leq 0} N_0(a, k) \otimes \mathcal{L}_{\chi_2})$ is nonzero. We know that $gr_W^{\leq 0} N_0(a, k)$ is negligible, and that its restriction to $\mathbb{G}_m \setminus \{1\}$ is lisse of rank 2, mixed of weights -1 and -3. Being negligible, it is, geometrically, a successive extension of Kummer objects $\mathcal{L}_\chi[1]$, so geometrically a direct sum of objects of the form $\mathcal{L}_\chi[1] \otimes Unip(n)$, each of which geometrically admits a quotient $\mathcal{L}_\chi[1]$. Thus $H_c^0(\mathbb{G}_m/\overline{k}, gr_W^{\leq 0} N_0(a, k) \otimes \mathcal{L}_{\overline{\chi}})$ is nonzero for each of the distinct χ which occur. In view of part (1) of the previous lemma, the only possibly nonvanishing $H_c^0(\mathbb{G}_m/\overline{k}, gr_W^{\leq 0} N_0(a, k) \otimes \mathcal{L}_\chi)$ has $\chi = \chi_2$. And as $gr_W^{\leq 0} N_0(a, k)$ is nonzero, being lisse of rank 2, there exist characters χ for which this group is nonzero. Thus $H_c^0(\mathbb{G}_m/\overline{k}, gr_W^{\leq 0} N_0(a, k) \otimes \mathcal{L}_{\chi_2})$ is nonzero, and χ_2 is the only character occurring in $gr_W^{\leq 0} N_0(a, k)$.

From the exact sequence, we get an inclusion

$$H_c^0(\mathbb{G}_m/\overline{k}, gr_W^{\leq 0} N_0(a, k) \otimes \mathcal{L}_{\chi_2}) \subset H_c^0(\mathbb{G}_m/\overline{k}, N_0(a, k) \otimes \mathcal{L}_{\chi_2}).$$

The first group is nonzero and mixed of weight < 0, while in the second, the space "weight < 0" is one-dimensional, and pure of weight -4. Therefore $H_c^0(\mathbb{G}_m/\overline{k}, gr_W^{\leq 0} N_0(a, k) \otimes \mathcal{L}_{\chi_2})$ is one-dimensional, and pure of weight -4, and the quotient

$$H_c^0(\mathbb{G}_m/\overline{k}, N_0(a, k) \otimes \mathcal{L}_{\chi_2})/H_c^0(\mathbb{G}_m/\overline{k}, gr_W^{\leq 0} N_0(a, k) \otimes \mathcal{L}_{\chi_2})$$

has rank 6 and is pure of weight 0.

We next claim that $H_c^1(\mathbb{G}_m/\overline{k}, gr_W^{\leq 0} N_0(a, k) \otimes \mathcal{L}_{\chi_2})$ is one-dimensional, and pure of weight 0. To see this, we consider the weight filtration on $gr_W^{\leq 0} N_0(a, k) \otimes \mathcal{L}_{\chi_2}$. The weights are -1 and -3, each of rank one. So we have a short exact sequence

$$0 \to V_{-4}[1] \to gr_W^{\leq 0} N_0(a, k) \otimes \mathcal{L}_{\chi_2} \to V_{-2}[1] \to 0,$$

where V_{-4} (resp. V_{-2}) is a geometrically constant sheaf of rank one which is pure of weight -4 (resp. -2). Take cohomology on $\mathbb{G}_m/\overline{k}$. Remember that $H_c^1(\mathbb{G}_m/\overline{k}, \overline{\mathbb{Q}_\ell}) = \overline{\mathbb{Q}_\ell}$ and $H_c^2(\mathbb{G}_m/\overline{k}, \overline{\mathbb{Q}_\ell}) = \overline{\mathbb{Q}_\ell}(-1)$. So we get a six term exact sequence

$$0 \to V_{-4} \to H_c^0(\mathbb{G}_m/\overline{k}, gr_W^{\leq 0} N_0(a, k) \otimes \mathcal{L}_{\chi_2}) \to V_{-2} \to V_{-4}(-1)$$
$$\to H_c^1(\mathbb{G}_m/\overline{k}, gr_W^{\leq 0} N_0(a, k) \otimes \mathcal{L}_{\chi_2}) \to V_{-2}(-1) \to 0.$$

The coboundary map $V_{-2} \to V_{-4}(-1)$ is a map between one-dimensional spaces, so it is either zero or an isomorphism. It cannot be zero, otherwise the group $H_c^0(\mathbb{G}_m/\overline{k}, gr_W^{\leq 0} N_0(a, k) \otimes \mathcal{L}_{\chi_2})$ would be two-dimensional. So it is an isomorphism, and hence we get isomorphisms

$$V_{-4} \cong H_c^0(\mathbb{G}_m/\overline{k}, gr_W^{\leq 0} N_0(a, k) \otimes \mathcal{L}_{\chi_2}),$$
$$H_c^1(\mathbb{G}_m/\overline{k}, gr_W^{\leq 0} N_0(a, k) \otimes \mathcal{L}_{\chi_2}) \cong V_{-2}(-1).$$

Therefore the group $H_c^0(\mathbb{G}_m/\overline{k}, N(a, k) \otimes \mathcal{L}_{\chi_2})$ is of dimension 7 and pure of weight 0, being an extension of $V_{-2}(-1)$ (rank 1, pure of weight 0) by

$$H_c^0(\mathbb{G}_m/\overline{k}, N_0(a, k) \otimes \mathcal{L}_{\chi_2})/H_c^0(\mathbb{G}_m/\overline{k}, gr_W^{\leq 0} N_0(a, k) \otimes \mathcal{L}_{\chi_2}),$$

(rank 6, pure of weight 0). $\qquad\square$

Lemma 27.6. *The object $N(a, k)$ is of the form $\mathcal{N}_a[1]$ for a sheaf $\mathcal{N}_a$. This sheaf $\mathcal{N}_a$ is lisse of rank 12 and pure of weight -1 on $\mathbb{G}_m \setminus \{1\}$, and for $j_1 : \mathbb{G}_m \setminus \{1\} \subset \mathbb{G}_m$ the inclusion, the adjunction map is an isomorphism*

$$\mathcal{N}_a \cong j_{1\star} j_1^\star \mathcal{N}_a.$$

The sheaf $\mathcal{N}_a$ is totally wild at both 0 and ∞.

Proof. The short exact sequence defining $N(a, k)$, namely

$$0 \to gr_W^{\leq 0}(N_0(a, k)) \to N_0(a, k) \to N(a, k) \to 0,$$

gives a six term exact sequence of ordinary cohomology sheaves

$$0 \to \mathcal{H}^{-1}(gr_W^{\leq 0}(N_0(a, k))) \to \mathcal{H}^{-1}(N_0(a, k)) \to \mathcal{H}^{-1}(N(a, k))$$
$$\to \mathcal{H}^0(gr_W^{\leq 0}(N_0(a, k))) \to \mathcal{H}^0(N_0(a, k)) \to \mathcal{H}^0(N(a, k)) \to 0.$$

The fourth and fifth terms, $\mathcal{H}^0(gr_W^{\leq 0}(N_0(a,k)))$ and $\mathcal{H}^0(N_0(a,k))$, both vanish (the fourth because $gr_W^{\leq 0}(N_0(a,k))$ is negligible, the fifth by part (1) of Lemma 27.1). Therefore the sixth term, $\mathcal{H}^0(N(a,k))$, vanishes. Thus $N(a,k) = \mathcal{N}_a[1]$ for $\mathcal{N}_a := \mathcal{H}^{-1}(N(a,k))$, and we have a short exact sequence of sheaves on $\mathbb{G}_m$

$$0 \to \mathcal{H}^{-1}(gr_W^{\leq 0}(N_0(a,k))) \to \mathcal{G}_a \to \mathcal{N}_a \to 0.$$

By Lemma 27.2, $\mathcal{N}_a|\mathbb{G}_m \setminus \{1\}$ is lisse of rank 12 and pure of weight -1. By Lemma 27.4, applied with any single character $\chi \neq \chi_2$, say with $\chi = \mathbb{1}$, the group $H_c^1(\mathbb{G}_m/\overline{k}, \mathcal{N}_a)$ is pure of weight 0, and all other H_c^i vanish. Exactly as in the proof of part (4) of Lemma 26.1, this purity implies that $\mathcal{N}_a$ is a middle extension. That $\mathcal{N}_a$ is totally wild at both 0 and ∞ is a restatement of the fact (Lemma 27.5) that $\mathcal{N}_a[1]$ has no bad characters. $\qquad\square$

Lemma 27.7. *The object $N(a,k)$ is its own Verdier dual, and it is isomorphic to its pullback by multiplicative inversion $t \mapsto 1/t$ on $\mathbb{G}_m$. As an object of $\mathcal{P}_{arith}$, $N(a,k)$ is self-dual.*

Proof. The third statement results from the first two. The invariance under $t \mapsto 1/t$ holds for the object $N_0(a,k)$ (by the same change of variable $z \mapsto z/t$, $t \mapsto t$) in the perverse sheaf

$$\mathcal{M} = \mathcal{M}(z,t) := (A^{-7})^{deg} \otimes \mathcal{H}_{3,1}(a/(z^2t)) \otimes Kl_2(z) \otimes Kl_2(zt)[2].$$

By the functoriality of the weight filtration, this invariance passes to $N(a,k) = gr_W^0(N(a,k))$. The perverse sheaf $\mathcal{M}$ is self-dual, so we get a cup product pairing

$$N_0(a,k) \times N_0(a,k) \to \overline{\mathbb{Q}_\ell}(1)[2].$$

Restricting to $\mathbb{G}_m \setminus \{1\}$, this is a pairing of lisse sheaves

$$\mathcal{G}_a \times \mathcal{G}_a \to \overline{\mathbb{Q}_\ell}(1).$$

By looking at the weights, we see that this pairing must annihilate $gr_W^{\leq -1}(\mathcal{G}_a)$ on either side, so it induces a pairing

$$\mathcal{N}_a \times \mathcal{N}_a \to \overline{\mathbb{Q}_\ell}(1).$$

Since this is a pairing of lisse sheaves on $\mathbb{G}_m \setminus \{1\}$, to check that it is a perfect pairing, it suffices to look at the stalk at a single point $t_0 \neq 1$. But at any such point, the stalk of N_a is the "pure of weight -1 part" of the stalk of $\mathcal{G}_a$. So this stalk is the image of H_c^1 in H^1 for the self-dual lisse sheaf

$$(A^{-7})^{deg} \otimes \mathcal{H}_{3,1}(a/(z^2t_0)) \otimes Kl_2(z) \otimes Kl_2(zt_0)[2],$$

on which the cup product pairing is known to be perfect. This shows that $N(a,k)|\mathbb{G}_m \setminus \{1\}$ is its own Verdier dual. By the previous lemma, $N(a,k)$ is the middle extension of its restriction to $\mathbb{G}_m \setminus \{1\}$, and one knows that middle extension commutes with formation of the Verdier dual. $\qquad\square$

Lemma 27.8. *We have the following results about the stalks $\mathcal{G}_{a1}$ and $\mathcal{N}_{a1}$ of $\mathcal{G}_a$ and $\mathcal{N}_a$ at the point 1.*

(1) *The stalk $\mathcal{G}_{a1}$ has rank 11. It has 8 Frobenius eigenvalues of weight -1, one Frobenius eigenvalue of weight -2, and two Frobenius eigenvalues of weight -4.*

(2) *The stalk $\mathcal{N}_{a1}$ has rank 9. It has 8 Frobenius eigenvalues of weight -1, and one Frobenius eigenvalue of weight -4.*

Proof. Denote by $j : \mathbb{G}_m \subset \mathbb{P}^1$ the inclusion. For the coefficient sheaf

$$\mathcal{K} := \mathcal{H}_{3,1}(a/(z^2)) \otimes Kl_2(z) \otimes Kl_2(z),$$

which is lisse on $\mathbb{G}_m$ and pure of weight 5, we have the short exact sequence with punctual quotient at 0 and ∞,

$$0 \to j_!\mathcal{K} \to j_*\mathcal{K} \to \mathcal{K}^{I(0)} \oplus \mathcal{K}^{I(\infty)} \to 0.$$

We have already proven in Lemma 27.1 that $H^2_c(\mathbb{G}_m/\overline{k}, \mathcal{K})$ vanishes. From the long exact cohomology we infer that $H^2(\mathbb{P}^1/\overline{k}, j_*\mathcal{K})$ vanishes, then by duality that $H^0(\mathbb{P}^1/\overline{k}, j_*\mathcal{K})$ vanishes. So we have a short exact sequence

$$0 \to \mathcal{K}^{I(0)} \oplus \mathcal{K}^{I(\infty)} \to H^1_c(\mathbb{G}_m/\overline{k}, \mathcal{K}) \to H^1(\mathbb{P}^1/\overline{k}, j_*\mathcal{K}) \to 0.$$

The third term is pure of weight 6 by Deligne's main theorem in Weil II [**De-Weil II**, 3.2.3]. So the drops in weight of the stalk $\mathcal{G}_{a1} = (A^{-7})^{deg} \otimes H^1_c(\mathbb{G}_m/\overline{k}, \mathcal{K})$ come from $\mathcal{K}^{I(0)} \oplus \mathcal{K}^{I(\infty)}$.

As we saw in the proof of Lemma 27.1, the isomorphism [**Ka-ClausCar**, 3.5]

$$Kl_2(z) \otimes Kl_2(z) \cong \overline{\mathbb{Q}_\ell}(-1) \oplus \mathcal{H}_{3,1}(4z) \otimes (A^{-1})^{deg}$$

gives a direct sum decomposition

$$\mathcal{K} \cong \mathcal{H}_{3,1}(a/(z^2))(-1) \bigoplus \mathcal{H}_{3,1}(a/(z^2)) \otimes \mathcal{H}_{3,1}(4z) \otimes (A^{-1})^{deg}.$$

The local monodromy of $\mathcal{H}_{3,1}(z)$ at 0 is $Unip(3)$, and at ∞ is

$$\mathcal{L}_{\chi_2} \oplus (\text{tot. wild, rank 2, both slopes } 1/2),$$

cf. [**Ka-ESDE**, 8.4.2]. So the local monodromy of $\mathcal{H}_{3,1}(a/(z^2))$ at 0 is $Unip(1) \oplus (\text{tot. wild, rank 2, both slopes } 1)$, and at ∞ is $Unip(3)$. Hence the local monodromy of $\mathcal{K}$ at 0 is the direct sum of

$$Unip(1) \oplus (\text{tot. wild, rank 2, both slopes } 1)$$

and

$$(Unip(1) \oplus (\text{tot. wild, rank 2, both slopes 1})) \otimes Unip(3).$$

The local monodromy at ∞ of $\mathcal{K}$ is the direct sum of $Unip(3)$ and

$$Unip(3) \otimes (\mathcal{L}_{\chi_2} \oplus (\text{tot. wild, rank 2, both slopes 1/2})).$$

Thus we have $Swan_0(\mathcal{K}) = 2 + 6 = 8$, $Swan_\infty(\mathcal{K}) = 3$, and hence $\mathcal{G}_{a1}$ has the asserted rank 11. The unipotent part of the local monodromy of $\mathcal{K}$ at 0 is $Unip(1) \oplus Unip(3)$, and the unipotent part of its local monodromy at ∞ is $Unip(3)$. These unipotent parts give the asserted weight drops of $1, 3, 3$ in $\mathcal{G}_{a1}$, cf. [**De-Weil II**, 1.6.14.2-3 and 1.8.4], [**Ka-GKM**, 7.0.7], thus proving the first assertion. The second assertion is immediate from the first, given that the kernel of the surjection of $\mathcal{G}_a$ onto $\mathcal{N}_a$ is lisse of rank 2, with weights -2 and -4. □

Lemma 27.9. *The local monodromy of $\mathcal{N}_a$ at 1 is $Unip(1)^8 \oplus Unip(4)$.*

Proof. From the weight drops given by part (2) of the previous lemma, we have a summand $Unip(1)^8 \oplus Unip(4)$. As the stalk has rank 12, there is room for no more. □

Lemma 27.10. *The sheaf $\mathcal{N}_a$ has $Swan_0(\mathcal{N}_a) = Swan_\infty(\mathcal{N}_a) = 2$.*

Proof. Identical to the proof of part (3) of Lemma 26.3. □

Lemma 27.11. *The object $N(a, k)$ is geometrically Lie-irreducible, and orthogonally self-dual.*

Proof. The proof is essentially identical to the proof of Lemma 26.4. One need only use the description of the local monodromy at 1 of $\mathcal{N}_a$, namely $Unip(1)^8 \oplus Unip(4)$, whereas Lemma 26.4 used that the local monodromy at 1 of $\mathcal{G}_a$ there was $Unip(1)^6 \oplus Unip(4)$. □

This concludes the construction of the objects $N(a, k)$ in odd characteristic.

CHAPTER 28

The Situation over $\mathbb{Z}$: Results

Suppose we are given an integer monic polynomial $f(x) \in \mathbb{Z}[x]$ of degree $n \geq 2$ which, over $\mathbb{C}$, is "weakly supermorse," meaning that it has n distinct roots in $\mathbb{C}$, its derivative $f'(x)$ has $n - 1$ distinct roots (the critical points) $\alpha_i \in \mathbb{C}$, and the $n - 1$ values $f(\alpha_i)$ (the critical values) are all distinct in $\mathbb{C}$. Denote by S the set of critical values. Suppose that S is not equal to any nontrivial multiplicative translate aS, for any $a \neq 1$ in $\mathbb{C}^\times$. It is standard that for all but finitely many primes p, the reduction mod p of f will satisfy all the hypotheses of Theorem 17.6. Let us say such a prime p is good for f.

Choose a prime ℓ, and a field isomorphism $\iota : \overline{\mathbb{Q}_\ell} \cong \mathbb{C}$. For each $p \neq \ell$ which is good for f, form the sheaf

$$\mathcal{F}_p := f_\star \overline{\mathbb{Q}_\ell}/\overline{\mathbb{Q}_\ell}|\mathbb{G}_m/\mathbb{F}_p,$$

and the corresponding object $N_p := \mathcal{F}_p(1/2)[1] \in \mathcal{P}_{arith}$ on $\mathbb{G}_m/\mathbb{F}_p$, which is pure of weight zero, geometrically irreducible, and has

$$G_{geom,N_p} = G_{arith,N_p} = GL(n-1).$$

We take the unitary group $U(n-1)$ as the compact form of $GL(n-1)$. For good p, the set of bad characters in this mod p situation is the set of characters of order dividing n. So for each good prime $p \neq \ell$, and for each character χ of $\mathbb{F}_p^\times$ with $\chi^n \neq \mathbb{1}$, we have the conjugacy class $\theta_{\mathbb{F}_p, \chi} \in U(n-1)^\#$.

Emanuel Kowalski asked if, as p grew, the sets of conjugacy classes

$$\{\theta_{\mathbb{F}_p, \chi}\}_{\chi \text{ char of } \mathbb{F}_p^\times, \ \chi^n \neq \mathbb{1}}$$

became equidistributed in the space $U(n-1)^\#$ of conjugacy classes of $U(n-1)$. We will show that this, and more general things of the same flavor, are true.

Here is the general setup. We fix a prime ℓ, a field isomorphism $\iota : \overline{\mathbb{Q}_\ell} \cong \mathbb{C}$, a reductive group G over $\overline{\mathbb{Q}_\ell}$, an integer $n \geq 1$, and a faithful n-dimensional representation of G, i.e., an inclusion $G \subset GL(n)$. We also fix a compact form K of $G(\mathbb{C})$.

We also fix a sequence of finite fields k_i, each of characteristic $\neq \ell$, whose cardinalities are nondecreasing and tend archimedeanly to ∞.

Thus for example the k_i could be successively higher degree extensions of a given prime field $\mathbb{F}_p$ with $p \neq \ell$, or the k_i could be a sequence of prime fields $\mathbb{F}_{p_i}$ with some increasing sequence of primes $p_i > \ell$, or it could be the sequence of residue fields of the closed points of some flat scheme of finite type over $\mathbb{Z}[1/\ell]$, e.g., the spectrum of some $\mathcal{O}_K[1/n\ell]$ for some number field K and some integer $n \neq 1$, or the sequence of finite fields k_i could be any amalgam of these example situations.

Theorem 28.1. *Suppose we are given, for each i, a form $\mathbb{G}_i/k_i$ of $\mathbb{G}_m/k_i$, and an arithmetically semisimple object N_i in $\mathcal{P}_{arith}$ on $\mathbb{G}_i/k_i$ which is ι-pure of weight zero, and of "dimension" n. Suppose that for every i we have*

$$G_{geom,N_i} = G_{arith,N_i} = G,$$

in such a way that the given n-dimensional representation of G, viewed as an n-dimensional representation of G_{arith,N_i}, corresponds to the object N_i. Suppose further that there exists a real number $C \geq n$ such that, for all i, we have both

$$gen.rk(N_i) \leq C,$$

and

$$\#Bad(N_i) \leq C.$$

Then the sets Θ_i of conjugacy classes in $K^\#$

$$\Theta_i := \{\theta_{k_i,\chi}\}_{\chi \text{ char of } \mathbb{G}_i(k_i),\ \chi \in Good(k_i,N_i)}$$

become equidistributed in $K^\#$ as $\#k_i \to \infty$.

Proof. Fix an irreducible nontrivial representation Λ of G. For each i, Λ corresponds to an object M_i in $<N_i>_{arith}$. We must show the large i limit of the averages

$$1/\#\Theta_i \sum_{\theta_{k_i,\chi} \in \Theta_i} \text{Trace}(\Lambda(\theta_{k_i,\chi}))$$

is zero.

Because $G \subset GL(V)$ is reductive, the irreducible nontrivial representation Λ occurs in some tensor space $V^{\otimes a} \otimes (V^\vee)^{\otimes b}$, for some pair (a,b) of nonnegative integers. Of course the pair (a,b) is not unique. Nonetheless, let us fix one such pair, say with $a + b$ minimal, for our given Λ. We will show that as soon as $\#k_i \geq (1+C)^2$, we have the explicit bound

$$|1/\#\Theta_i \sum_{\theta_{k_i,\chi} \in \Theta_i} \text{Trace}(\Lambda(\theta_{k_i,\chi}))| \leq 2(a+b+1)C^{a+b}/\sqrt{\#k_i}.$$

Recall from Remark 7.5 the explicit estimate (applied here with $N = N_i$ and $E = k_i$, and valid as soon as $C \leq \sqrt{\#k_i} - 1$)

$$|(1/\#Good(k_i, N_i)) \sum_{\rho \in Good(k_i, N_i)} \text{Trace}(\Lambda(\theta_{k_i, \rho})|$$

$$\leq 2(gen.rk(M_i) + \text{``dim''}(M_i))/\sqrt{\#k_i}.$$

Since M_i is a direct summand of the tensor product in the Tannakian sense $N_i^{\otimes a} \otimes (N_i^\vee)^{\otimes b}$, we certainly have

$$\text{``dim''}(M_i) \leq \text{``dim''}(N_i^{\otimes a} \otimes (N_i^\vee)^{\otimes b}) = n^{a+b} \leq C^{a+b}$$

and

$$gen.rk(M_i) \leq gen.rk(N_i^{\otimes a} \otimes (N_i^\vee)^{\otimes b}).$$

So it is sufficient to establish the inequality

$$gen.rk(N_i^{\otimes a} \otimes (N_i^\vee)^{\otimes b}) \leq (a + b)C^{a+b}.$$

This in turn is a consequence of the inequality

$$gen.rk(N_i^{\otimes a} \otimes (N_i^\vee)^{\otimes b}) \leq (a + b)(\text{``dim''}(N_i))^{a+b-1}gen.rk(N_i).$$

This inequality results, by induction on $a+b$, from the following general inequality, which will be proven below. Over any algebraically closed field of characterstic $p \neq \ell$, given two geometrically semisimple objects N and M in $\mathcal{P}_{geom}$, we claim that we have the inequality

$$gen.rk(N \star_{mid} M) \leq \text{``dim''}(N)gen.rk(M) + gen.rk(N)\text{``dim''}(M).$$

Indeed, if we grant this general inequality, then if $a \geq 1$ we get

$$gen.rk(N_i^{\otimes a} \otimes (N_i^\vee)^{\otimes b})$$

$$= gen.rk(N \star_{mid} (N_i^{\otimes a-1} \otimes (N_i^\vee)^{\otimes b}))$$

$$\leq \text{``dim''}(N_i)gen.rk(N_i^{\otimes a-1} \otimes (N_i^\vee)^{\otimes b}) + gen.rk(N_i)\text{``dim''}(N_i^{\otimes a-1} \otimes (N_i^\vee)^{\otimes b}).$$

By induction on $a + b$, we have

$$gen.rk(N_i^{\otimes a-1} \otimes (N_i^\vee)^{\otimes b}) \leq (a + b - 1)\text{``dim''}(N_i))^{a+b-2}gen.rk(N_i),$$

so we get the inequality

$$gen.rk(N_i^{\otimes a} \otimes (N_i^\vee)^{\otimes b}) \leq (a + b)\text{``dim''}(N_i))^{a+b-1}gen.rk(N_i)$$

as asserted. If $a = 0$, we repeat this argument, "factoring out" one $N_i^\vee$ and doing induction on b. $\qquad\square$

We now establish the inequality used in the proof of Theorem 28.1. Because the function $N \mapsto$ "dim"(N) is multiplicative, i.e., "dim"$(N \star_{mid} M) =$ "dim"(N)"dim"(M), it is as though the inequality asserts a (sub)product formula for "differentiation," where the function $N \mapsto gen.rk(N)$ plays the role of the "derivative" of the "dim" function.

Theorem 28.2. *Over an algebraically closed field $\overline{k}$ of characterstic $p \neq \ell$, given two objects geometrically semisimple N and M in $\mathcal{P}_{geom}$, we have the inequality*

$$gen.rk(N \star_{mid} M) \leq \text{``dim''}(N)gen.rk(M) + gen.rk(N)\text{``dim''}(M).$$

Proof. Since all the terms in the asserted inequality are bilinear in the arguments N and M, we reduce immediately to the case where N and M are both geometrically irreducible.

If either N or M is punctual, say $N = \delta_a$, then $N \star_{mid} M$ is just the multiplicative translate $[x \mapsto ax]_\star M = [x \mapsto x/a]^\star M$ of M by a, so has the same generic rank as M. But "dim"$(N) = 1$, and $gen.rk(N) = 0$, so in this case we have equality.

Suppose now that our two objects in $\mathcal{P}_{geom}$ are each (geometrically irreducible, but we will not use this) middle extension sheaves placed in degree -1, say $N = \mathcal{F}[1]$ and $M = \mathcal{G}[1]$. In this case, we first use the fact that $N \star_{mid} M$ is a quotient of $N \star_! M$, so we have the trivial inequality

$$gen.rk(N \star_{mid} M) \leq gen.rk(N \star_! M).$$

So it suffices to show the inequality

$$gen.rk(N \star_! M) \leq \text{``dim''}(N)gen.rk(M) + gen.rk(N)\text{``dim''}(M).$$

Over a dense open set $U \subset \mathbb{G}_m/\overline{k}$, $N \star_! M|U$ is of the form $\mathcal{H}[1]$ for the lisse sheaf on U whose stalk at a point $a \in U$ is the cohomology group $H^1_c(\mathbb{G}_m/\overline{k}, \mathcal{F} \otimes [x \mapsto a/x]^\star \mathcal{G})$, and this H^1_c is the only nonvanishing cohomology group. So for any $a \in U(\overline{k})$, we have

$$gen.rk(N \star_! M) = -\chi(\mathbb{G}_m/\overline{k}, \mathcal{F} \otimes [x \mapsto a/x]^\star \mathcal{G}).$$

Shrinking U if necessary, we may further assume that for any point $a \in U(\overline{k})$, the two middle extension sheaves $\mathcal{F}$ and $[x \mapsto a/x]^\star \mathcal{G}$ on $\mathbb{G}_m$ have disjoint sets of ramification in $\mathbb{G}_m$. Choose one point $a \in U(\overline{k})$, and define

$$\mathcal{K} := [x \mapsto a/x]^\star \mathcal{G}.$$

Then

$$\text{``dim''}(N) = -\chi(\mathbb{G}_m/\overline{k}, \mathcal{F}),$$
$$\text{``dim''}(M) = \text{``dim''}([x \mapsto a/x]^\star M) = -\chi(\mathbb{G}_m/\overline{k}, \mathcal{K}),$$

and
$$gen.rk(N \star_! M) = -\chi(\mathbb{G}_m/\overline{k}, \mathcal{F} \otimes \mathcal{K}).$$
So it suffices to prove that for two middle extension sheaves $\mathcal{F}$ and $\mathcal{K}$ on $\mathbb{G}_m/\overline{k}$ with disjoint ramification, we have

$$-\chi(\mathbb{G}_m/\overline{k}, \mathcal{F} \otimes \mathcal{K}) \leq -gen.rk(\mathcal{F})\chi(\mathbb{G}_m/\overline{k}, \mathcal{K}) - gen.rk(\mathcal{K})\chi(\mathbb{G}_m/\overline{k}, \mathcal{F}).$$

Now for any sheaf $\mathcal{A}$ on $\mathbb{G}_m/\overline{k}$, with ramification set $Ram(\mathcal{A}) \subset \mathbb{G}_m$, the Euler-Poincaré formula [**Ray**] gives

$$-\chi(\mathbb{G}_m/\overline{k}, \mathcal{A}) = Swan_0(\mathcal{A}) + Swan_\infty(\mathcal{A}) + \sum_{r \in Ram(\mathcal{A})} (drop_r(\mathcal{A}) + Swan_r(\mathcal{A})).$$

Taking $\mathcal{A}$ to be $\mathcal{F} \otimes \mathcal{K}$, and denoting by S and T the disjoint ramification sets of $\mathcal{F}$ and of $\mathcal{K}$, so that $S \cup T$ is the ramification set of $\mathcal{A}$, we have (precisely because the ramification is disjoint)

$$gen.rk(N \star_! M) = Swan_0(\mathcal{F} \otimes \mathcal{K}) + Swan_\infty(\mathcal{F} \otimes \mathcal{K})$$
$$+ \sum_{s \in S}(drop_s(\mathcal{F}) + Swan_s(\mathcal{F}))gen.rk(\mathcal{K})$$
$$+ \sum_{t \in T}(drop_t(\mathcal{K}) + Swan_t(\mathcal{K}))gen.rk(\mathcal{F}).$$

Let us admit for a moment the following two inequalities.

$$Swan_0(\mathcal{F} \otimes \mathcal{K}) \leq Swan_0(\mathcal{F})gen.rk(\mathcal{K}) + Swan_0(\mathcal{K})gen.rk(\mathcal{F})$$

and

$$Swan_\infty(\mathcal{F} \otimes \mathcal{K}) \leq Swan_\infty(\mathcal{F})gen.rk(\mathcal{K}) + Swan_\infty(\mathcal{K})gen.rk(\mathcal{F}).$$

Then we have the inequality

$$gen.rk(N \star_! M) \leq Swan_0(\mathcal{F})gen.rk(\mathcal{K}) + Swan_0(\mathcal{K})gen.rk(\mathcal{F})$$
$$+ Swan_\infty(\mathcal{F})gen.rk(\mathcal{K}) + Swan_\infty(\mathcal{K})gen.rk(\mathcal{F})$$
$$+ \sum_{s \in S}(drop_s(\mathcal{F}) + Swan_s(\mathcal{F}))gen.rk(\mathcal{K})$$
$$+ \sum_{t \in T}(drop_t(\mathcal{K}) + Swan_t(\mathcal{K}))gen.rk(\mathcal{F}).$$

Factoring out the $gen.rk(\mathcal{F})$ and $gen.rk(\mathcal{K})$ terms, we see that this is precisely the inequality

$$gen.rk(N \star_! M) \leq \text{``dim''}(M)gen.rk(N) + \text{``dim''}(N)gen.rk(M).$$

So it remains to show that for two $\overline{\mathbb{Q}_\ell}$-representations A and B of $I(0)$, we have

$$Swan_0(A \otimes B) \leq Swan_0(A)\text{rk}(B) + Swan_0(B)\text{rk}(A),$$

(and similarly for two $\overline{\mathbb{Q}_\ell}$-representations of $I(\infty)$).

Again by bilinearity, we may use the slope decomposition [**Ka-GKM**, Chapter 1] to reduce to the case where A and B each have only a single slope, say λ is the unique slope of A and μ is the unique slope of B, with say $\lambda \le \mu$. Then $I(0)^{\mu+}$ acts trivially on both A and B, so also trivially on $A \otimes B$. Therefore all slopes of $A \otimes B$ are at most μ, so we have

$$Swan_0(A \otimes B) \le \mu\mathrm{rk}(A \otimes B) = \mathrm{rk}(A)\mu\mathrm{rk}(B)$$
$$= Swan_0(B)\mathrm{rk}(A) \le Swan_0(B)\mathrm{rk}(A) + Swan_0(A)\mathrm{rk}(B).$$

$\square$

Remark 28.3. How can one apply Theorem 28.1? Take for G one of the groups $SL(n)$, $GL(n)$, $Sp(n)$, or a self-product of one of these groups, or $O(2n)$. In Chapters 14, 15, 17, 18, 20, 21, 22, 23 and 24 we give various theorems which assert that a specific[1] N_i over a finite field k_i has $G_{geom,N_i} = G_{arith,N_i} = G$. Fix one such G, and choose one of the theorems in the relevant chapter. In all of these theorems, it happens that both the generic rank and the number of bad characters of the example object are bounded in terms of the "dimension" of that object. Given a sequence of finite fields k_i, each of characteristic $\neq \ell$, whose cardinalities are nondecreasing and tend archimedeanly to ∞, we have only to pick, for each i, an instance N_i over k_i of the chosen theorem, to have data to which Theorem 28.1 applies. Thus for example if we invoke Corollary 20.2 for some given value of $r \ge 1$, we must first "throw way" those k_i of cardinality $\le r$, and for each of the remaining ones choose both a nontrivial additive character ψ_i of k_i and r distinct multiplicative characters of $k_i^\times$. But there is no "compatiblity" of any kind required in these choices as k_i varies. Similarly, one way to invoke Theorem 17.6 is to start, as we did this chapter, with an integer monic polynomial $f(x) \in \mathbb{Z}[x]$ of degree $n \ge 2$ which, over $\mathbb{C}$, is "weakly supermorse," and whose set S of critical values is not equal to any nontrivial multiplicative translate aS, for any $a \neq 1$ in $\mathbb{C}^\times$. Then modulo sufficiently large primes p, the reduction mod p of f will satisfy all the hypotheses of Theorem 17.6. **However**, we could equally well invoke Theorem 17.6 by choosing separately for each (large) prime p a degree n monic polynomial $f_p(x) \in \mathbb{F}_p[x]$ to which the theorem applies.

Remark 28.4. In Chapter 19, which treats the orthogonal group, we only give situations where $G_{geom,N_i} = SO(n)$ and $G_{arith,N_i} \subset O(n)$, but no specific situations where we know that $G_{geom,N_i} = G_{arith,N_i} = SO(n)$. But as we observed in Remark 19.13, in any such situation we

---

[1]The results of Chapters 25-27, devoted to G_2, fail to provide specific examples.

will always achieve $G_{geom,N_i} = G_{arith,N_i} = SO(n)$ after a quadratic extension of the ground field. So here we have the somewhat less optimal situation that if we pick one of the explicit example theorems of Chapter 19, then given our sequence of finite fields k_i, we pick for each an instance of that theorem, but only over the sequence of the quadratic extensions of the given k_i will we have data to which we can apply Theorem 28.1 with G taken to be $SO(n)$.

CHAPTER 29

The Situation over ℤ: Questions

There is another sense in which we might ask about "situations over ℤ," namely we might try to mimic the setting of a theorem of Pink [**Ka-ESDE**, 8.18.2] about how "usual" (geometric) monodromy groups vary in a family. There the situation is that we are given a normal noetherian connected scheme S, a smooth X/S with geometrically connected fibres, and a lisse $\overline{\mathbb{Q}_\ell}$-sheaf $\mathcal{F}$ on X of rank $n \geq 1$. For each geometric point s in S, we have the restriction $\mathcal{F}_s$ of $\mathcal{F}$ to the fibre X_s. Pick a geometric point x_s in X_s. Then for each s in S we have the closed subgroup $\Gamma(s) \subset GL(n, \overline{\mathbb{Q}_\ell})$ which is the image of $\pi_1(X_s, x_s)$ in the representation corresponding to $\mathcal{F}_s$. The assertion is that these groups $\Gamma(s)$ are, up to $GL(n)$-conjugacy, constant on a dense open set of S, and that they decrease under specialization.

This result leads naturally to the following question. Suppose we are given a normal noetherian connected scheme S which is of finite type over $\mathbb{Z}[1/\ell]$, an object N in the derived category $D_c^b((\mathbb{G}_m)_S, \overline{\mathbb{Q}_\ell})$, and an integer $n \geq 1$. We make the following assumptions on this data.

(1) For every geometric point $\overline{s}$ of S, the restriction, $N_{\overline{s}}$, of N to the fibre over $\overline{s}$ is perverse and lies in $\mathcal{P}$.

(2) The formation of the Verdier dual, $D_{(\mathbb{G}_m)_S/S}(N)$, commutes with arbitrary change of base on S to a good scheme S' (a condition which is always satisfied after we shrink S to a dense open set of itself [**Ka-Lau**, 1.1.7]).

(3) For every finite field k and for every k-valued point $s \in S(k)$, the restriction $N_{k,s}$ of N to the $\mathbb{G}_m/k$ which is the fibre of $(\mathbb{G}_m)_S$ over s is perverse, lies in $\mathcal{P}_{arith}$, has "dimension" n, and is geometrically semisimple. [This last condition holds if, for example, each $N_{k,s}$ is pure of some integer weight w.]

The first two conditions say that N is "perverse relative to S" in the sense of [**Ka-Lau**, 1.2.1], and in addition satisfies $\mathcal{P}$ on each geometric fibre. For η the generic point of S, and $\overline{\eta}$ a geometric point lying over it, we have the perverse object $N_{\overline{\eta}}$ on $\mathbb{G}_m$ over an algebraically closed field $\kappa(\overline{\eta})$, which lies in $\mathcal{P}$. The results of Gabber-Loeser developed in their seminal paper, especially [**Ga-Loe**, 3.7.2,3.7.5], are stated when the ground field has strictly positive characteristic, but remain valid in the case of characteristic zero, cf. [**Ga-Loe**, lines 22-25 on page 505]. So whatever the characteristic of $\kappa(\overline{\eta})$, we may speak of the Tannakian group $G_{geom,N_{\overline{\eta}}}$. Using the construcibility of the Euler characteristic on fibres, the third condition implies that $N_{\overline{\eta}}$ has "dimension" n. Hence we have an inclusion $G_{geom,N_{\overline{\eta}}} \subset GL(n)$, well-defined up to $GL(n)$-conjugacy.

On the other hand, for every finite field k and for every k-valued point $s \in S(k)$, we have the Tannakian group $G_{geom,N_{k,s}} \subset GL(n)$. The first natural question is whether this group is always conjugate, in $GL(n)$, to a **subgroup** of $G_{geom,N_{\overline{\eta}}}$. We cannot expect equality of these groups, even if we are willing to shrink S to an open dense subset of itself, as the following example shows. [See [**Ka-ESDE**, 2.4.1, 2.4.4] for a discussion of the analogous phenomenon for differential galois groups.]

Take for S the spectrum of $\mathbb{Z}[1/\ell]$. Then ℓ is a global section of $(\mathbb{G}_m)_S$, so we can speak of the delta sheaf δ_ℓ. The finite fields k for which $S(k)$ is nonempty are precisely those of characteristic $p \neq \ell$, and for each such k there is a unique point s in $S(k)$. For each such point s, the Tannakian group $G_{geom,(\delta_\ell)_{k,s}}$ is the finite group $\mu_{N(\ell,p)} \subset GL(1)$ of roots of unity of order $N(\ell,p) :=$ the multiplicative order of ℓ mod p. As this order, for fixed ℓ and variable p, is unbounded (otherwise ℓ would itself be a root of unity in $\mathbb{Q}$), these Tannakian groups do not become constant, no matter how large the finite set of primes p we invert. Nor do they ever become the entire group $GL(1)$, which is the value of $G_{geom,N_{\overline{\eta}}}$ (exactly because ℓ has infinite multiplicative order in $\mathbb{Z}[1/\ell]^\times$).

So two plausible questions in this context are the following.

(Q1) Is it true that for every finite field k and for every k-valued point $s \in S(k)$, the Tannakian group $G_{geom,N_{k,s}}$ is conjugate in $GL(n)$ to a subgroup of $G_{geom,N_{\overline{\eta}}}$?

(Q2) After possibly shrinking S to an open dense subset $U \subset S$, is it true that for every finite field k and for every k-valued point $u \in U(k)$, the derived group (commutator subgroup) of the identity component of $G_{geom,N_{k,u}}$, $((G_{geom,N_{k,u}})^0)^{der}$, is conjugate in $GL(n)$ to the derived group of the identity component of $G_{geom,N_{\overline{\eta}}}$, $((G_{geom,N_{\overline{\eta}}})^0)^{der}$?

Another natural question, involving only finite field fibres, is this.

(Q3) Suppose given an object N as above, and a reductive group $G \subset GL(n)$. Suppose there is a dense open set $U \subset S$ such that for every finite field k and every k-valued point $u \in U(k)$,

$G_{geom,N_{k,u}}$ is conjugate in $GL(n)$ to the given group G. Is it then true that for every finite field k and for every k-valued point $s \in (S \setminus U)(k)$, $G_{geom,N_{k,s}}$ is conjugate in $GL(n)$ to a subgroup of the given group G?

Here is an example, based on Theorem 18.7, where this last question has an affirmative answer.

Example 29.1. Fix an even integer $2g \geq 4$, and consider the one parameter ("a") family of palindromic polynomials

$$f_a(x) := x^{2g} + ax^{2g-1} + ax + 1.$$

Denote by $\Delta(a) \in \mathbb{Z}[a]$ the discriminant of f_a. Its constant term is invertible in $\mathbb{Z}[1/2g]$ (because $f_0(x) = x^{2g} + 1$ has $2g$ distinct roots in any field in which $2g$ in invertible). Take for S the spectrum of the ring $\mathbb{Z}[1/2g\ell][a][1/\Delta(a)]$. On $(\mathbb{G}_m)_S$, we have the polynomial function $f_a(x)$, and the dense open set $j : (\mathbb{G}_m)_S[1/f_a(x)] \subset (\mathbb{G}_m)_S$. We take for N the object

$$N := j_! \mathcal{L}_{\chi_2(f_a(x))}[1].$$

Then on each geometric fibre of $(\mathbb{G}_m)_S/S$, the object $N(1/2)$ is perverse, has $\mathcal{P}$, is pure of weight zero and is symplectically self-dual of "dimension" $2g$ ("dimension" $2g$ because we inverted Δ). On any fibre over the open set $S[1/a]$ where a is invertible, f_a has an x-term, so is not a polynomial in x^d for any $d \geq 2$. So by Theorem 18.7, over $U := S[1/a]$, we are in the situation of this last question, with $G = Sp(2g)$.

For points in $S \setminus U = Spec(\mathbb{Z}[1/2g\ell])$, i.e., points where $a = 0$, we are looking at

$$N_0 := j_! \mathcal{L}_{\chi_2(x^{2g}+1)}[1].$$

Over any field in which $2g\ell$ is invertible, $N_0(1/2)$ remains symplectic, but it is no longer Lie-irreducible, cf. Corollary 8.3. So its G_{geom} is a subgroup of $Sp(2g)$, but it is no longer the entire symplectic group.

We can be more precise about what its G_{geom} is. The key observation is that $\mathcal{L}_{\chi_2(x^{2g}+1)}$ is the Kummer pullback $[2g]^*(\mathcal{L}_{\chi_2(x+1)})$ of $\mathcal{L}_{\chi_2(x+1)}$. Its direct image by $[2g]$ is the direct sum

$$[2g]_*(\mathcal{L}_{\chi_2(x^{2g}+1)}) = [2g]_*[2g]^*(\mathcal{L}_{\chi_2(x+1)}) = \oplus_i \mathcal{L}_{\chi_2(x+1)} \otimes \mathcal{L}_{\rho_i}$$

over the $2g$ multiplicative characters ρ_i of order dividing $2g$. The dual of $j_! \mathcal{L}_{\chi_2(x+1)}[1](1/2)$ is $j_! \mathcal{L}_{\chi_2((1/x)+1)}[1](1/2) = j_! \mathcal{L}_{\chi_2(x+1)} \otimes \mathcal{L}_{\chi_2}[1](1/2)$. So for each ρ_i, the dual of $j_! \mathcal{L}_{\chi_2(x+1)} \otimes \mathcal{L}_{\rho_i}[1](1/2)$ is $j_! \mathcal{L}_{\chi_2(x+1)} \otimes \mathcal{L}_{\rho_i\chi_2}[1](1/2)$. Now apply Corollary 20.3, with $a := -1$, $\Lambda := \chi_2$, and a choice of g

among the ρ_i which picks one out of each pair $(\rho_i, \chi_2 \rho_i)$. Then the partial direct sum

$$\oplus_{\text{chosen } \rho_i} j_! \mathcal{L}_{\chi_2(x+1)} \otimes \mathcal{L}_{\rho_i}[1](1/2)$$

has its G_{geom} a g-dimensional torus, with the unchosen ρ_i terms being the inverse characters. In other words, the G_{geom} of the entire direct sum

$$[2g]_* N_0(1/2) = \oplus_{\text{all } \rho_i} j_! \mathcal{L}_{\chi_2(x+1)} \otimes \mathcal{L}_{\rho_i}[1](1/2)$$

is the maximal torus of $Sp(2g)$. So by Theorems 8.1 and 8.2, we see that for $N_0 := j_! \mathcal{L}_{\chi_2(x^{2g}+1)}[1]$ itself, the identity component of its G_{geom} is the maximal torus of $Sp(2g)$.

Here is another example, exhibiting more extensive and interesting specialization behavior.

Example 29.2. Again fix an even integer $2g \geq 4$, and consider now the two parameter family of polynomials

$$f_{a,b}(x) = x^{2g} + ax^{2g-1} + bx + 1.$$

We denote by $\Delta(a, b) \in \mathbb{Z}[a, b]$ its discriminant. Just as in the previous example, its constant term is invertible in $\mathbb{Z}[1/2g]$. We now take for S the spectrum of the ring $\mathbb{Z}[1/2g\ell][a, b][1/\Delta(a, b)]$, and for N the object

$$N := j_! \mathcal{L}_{\chi_2(f_{a,b}(x))}[1].$$

Using now Theorem 23.2, elementary computation shows that over the locus where $a^{2g} - b^{2g}$ is invertible, we have $G_{geom} = SL(2g)$.

The locus where ab is invertible and $a^{2g} - b^{2g}$ is not invertible is, set theoretically, the disjoint union over divisors d of $2g$, of the sets where $\zeta := a/b$ is a primitive d'th root of unity. As we will see below, the question of whether or not d divides g, i.e., whether $\zeta^g = 1$ or $\zeta^g = -1$, has a huge effect on what G_{geom} turns out to be.

Suppose first that $\zeta^g = 1$, i.e., that $d|g$. We claim that on the locus $a = \zeta b$, b invertible, G_{geom} is the group $GSp_d(2g) := \mu_{2d} Sp(2g)$ of **symplectic** similitudes with multiplicator of order dividing d. Indeed, if we choose a square root η of ζ of order $2d$ (both choices have order $2d$ if d is even, just one choice does if d is odd), then $f_{a,b}(\eta x)$ is palindromic, and we are dealing with its multiplicative translate by η. Then $\delta_{1/\eta}$ "is" a character of order $2d$. By Theorem 18.7, the direct sum

$$j_! \mathcal{L}_{\chi_2(f_{a,b}(\eta x))}[1] \oplus \delta_\eta$$

has its G_{geom} a subgroup of the product group $Sp(2g) \times \mu_{2d}$, which maps onto each factor. As $Sp(2g)$ has no nontrivial quotients, Goursat's lemma shows that this direct sum has its G_{geom} the product group, and hence our N has its G_{geom} equal to $\mu_{2d} Sp(2g)$. [If d is odd and we chose η of order d, we would find that G_{geom} is $\mu_d Sp(2g)$, which for d odd is also group $\mu_{2d} Sp(2g)$.]

Suppose next that $\zeta^g = -1$, i.e., that d does not divide g. We claim that on the locus $a = \zeta b$, b invertible, G_{geom} is $SL(2g) \cap (\mu_{2d} O(2g))$, the group of **orthogonal** similitudes with multiplicator of order d and determinant 1. Indeed, for either choice of η with $\eta^2 = \zeta$ (here d is even, and so both choices of η have order $2d$), $f_{a,b}(\eta x)$ is **antipalindromic**, and we are again dealing with its multiplicative translate by η. Here, however, there is an additional subtlety. By Theorem 24.1, the direct sum

$$j_! \mathcal{L}_{\chi_2(f_{a,b}(\eta x))}[1] \oplus \delta_\eta$$

has its G_{geom} a subgroup of the product group $O(2g) \times \mu_{2d}$, which maps onto each factor. By Goursat's lemma, this G_{geom} is the (inverse image in the product of the) graph of an isomorphism of a quotient of $O(2g)$ with a quotient of μ_{2d}. The only nontrivial abelian quotient of $O(2g)$ is ± 1, via the determinant, since $SO(2g)$ is, for $2g \geq 4$, its own commutator subgroup. The only ± 1 quotient of μ_{2d} is given by the d'th power map. Because d divides $2g$ but does not divide g, the ratio $2g/d$ is odd. So we may also describe the only ± 1 quotient of μ_{2d} as given by the $2g$'th power map. So for the direct sum above, its G_{geom} is either the full product $\mu_{2d} \times O(2g)$ or it is the subgroup consisting of those elements $(\gamma \in \mu_{2d}, A \in O(2g))$ satisfying $det(A) = \gamma^{2g}$. So for the object N, its G_{geom} is either $\mu_{2d} O(2g)$ or it is the intersection of that group with $SL(2g)$. From Theorem 23.2, we know that $G_{geom,N}$ lies in $SL(2g)$, and this rules out the $\mu_{2d} O(2g)$ possibility.

On the locus $a = b = 0$, the identity component of G_{geom} is, as we have seen in Example 29.1, the maximal torus of $Sp(2g)$.

CHAPTER 30

Appendix: Deligne's Fibre Functor

In this appendix, we prove Theorem 3.1, i.e., we show that $N \mapsto \omega(N) := H^0(\mathbb{A}^1/\overline{k}, j_{0!}N)$ is a fibre functor on the Tannakian category $\mathcal{P}_{geom}$ of those perverse sheaves on $\mathbb{G}_m/\overline{k}$ satisfying $\mathcal{P}$, under middle convolution. Throughout this appendix, we work entirely over $\overline{k}$, explicit mention of which we will omit. Thus we will write $\omega(N)$ simply as $H^0(\mathbb{A}^1, j_{0!}N)$. And when we wish to emphasize the roles of both 0 and ∞ in its definition, we will write it as $\omega(N) := H^0(\mathbb{P}^1, Rj_{\infty\star}j_{0!}N) = H^0(\mathbb{P}^1, j_{0!}Rj_{\infty\star}N)$.

It will be convenient to define, for **any** object $M \in D_c^b(\mathbb{G}_m, \overline{\mathbb{Q}_\ell})$,

$$\omega(M) := H^\star(\mathbb{P}^1, j_{0!}Rj_{\infty\star}M) := \oplus_{i \in \mathbb{Z}} H^i(\mathbb{P}^1, j_{0!}Rj_{\infty\star}M).$$

Lemma 30.1. *If M is perverse, then $H^i(\mathbb{P}^1, j_{0!}Rj_{\infty\star}M) = 0$ for $i \neq 0$, and $\dim H^0(\mathbb{P}^1, j_{0!}Rj_{\infty\star}M) = \chi(\mathbb{G}_m, M)(= \chi_c(\mathbb{G}_m, M))$. The functor $M \mapsto \omega(M)$ is an exact, faithful functor from the category $Perv/Neg$ to the category of finite dimensional $\overline{\mathbb{Q}_\ell}$-vector spaces.*

Proof. We can also write $H^\star(\mathbb{P}^1, j_{0!}Rj_{\infty\star}M)$ as $H^\star(\mathbb{A}^1, j_{0!}M)$. Thus the first assertion was proven in Lemma 3.4. The dimension formula shows that $\omega(M) = 0$ if and only if M is negligible. The (not very) long exact cohomology sequence shows that ω is exact on $Perv/Neg$, and kills precisely the negligible objects. $\qquad\square$

Suppose now that K and N are both in $\mathcal{P}_{geom}$. Then both $K \star_! N$ and $K \star_\star N$ lie are perverse, and $K\star_{mid}$ is, by definition, the image of the "forget supports" map $K \star_! N \to K \star_\star N$. By Gabber-Loeser [**Ga-Loe**, 3.6.4], both the kernel and cokernel of this map are negligible perverse sheaves, i.e., with $\chi = 0$. Hence we get

Lemma 30.2. *For K and N in $\mathcal{P}_{geom}$, under the natural maps*

$$K \star_! N \twoheadrightarrow K \star_{mid} N \hookrightarrow K \star_\star N,$$

the induced maps on ω's are isomorphisms

$$\omega(K \star_! N) \cong \omega(K \star_{mid} N) \cong \omega(K \star_\star N).$$

Lemma 30.3. *For M perverse, with $M^\vee := [x \mapsto 1/x]^\star DM$, $\omega(M^\vee)$ is the linear dual of $\omega(M)$.*

Proof. Indeed,

$$\omega(M^\vee) := H^0(\mathbb{P}^1, j_{0!}Rj_{\infty\star}[x \mapsto 1/x]^\star DM) = H^0(\mathbb{P}^1, [x \mapsto 1/x]^\star j_{\infty!}Rj_{0\star}DM$$

$$= H^0(\mathbb{P}^1, [x \mapsto 1/x]^\star D(Rj_{\infty\star}j_{0!}M)) \overset{[x \mapsto 1/x]^\star}{\cong} H^0(\mathbb{P}^1, D(Rj_{\infty\star}j_{0!}M)).$$

$\square$

Lemma 30.4. *Given two objects K and N in $\mathcal{P}_{geom}$, we have the formula*

$$\dim(\omega(K) \otimes \omega(N)) = \dim \omega(K \star_! N),$$

i.e., we have

$$\chi(\mathbb{G}_m, K) \times \chi(\mathbb{G}_m, N) = \chi(\mathbb{G}_m, K \star_! N).$$

Proof. Indeed, for $mult : \mathbb{G}_m \times \mathbb{G}_m \to \mathbb{G}_m$ the multiplication map, we have $K \star_! N = R(mult)_!(K \boxtimes N)$. Thus

$$\chi(\mathbb{G}_m, K \star_! N) = \chi(\mathbb{G}_m, R(mult)_!(K \boxtimes N))$$

$$\overset{\text{Leray}}{=} \chi(\mathbb{G}_m \times \mathbb{G}_m, K \boxtimes N) \overset{\text{Kunneth}}{=} \chi(\mathbb{G}_m, K) \times \chi(\mathbb{G}_m, N).$$

$\square$

We now turn to the proof of Theorem 3.1. Given two objects K and N in $\mathcal{P}_{geom}$, we will define bifunctorial maps

$$\omega(K) \otimes \omega(N) \to \omega(K \star_! N) \cong \omega(K \star_{mid} N) \cong \omega(K \star_\star N) \to \omega(K) \otimes \omega(N),$$

and show that their composite is the identity on $\omega(K) \otimes \omega(N)$. This is all we need: each of the five terms has dimension $\chi(\mathbb{G}_m, K) \times \chi(\mathbb{G}_m, N)$, so once the composite map is the identity, then the first map is injective and hence, for dimension reasons, an isomorphism. This isomorphism gives the required bifunctorial isomorphism

$$\omega(K) \otimes \omega(N) \cong \omega(K \star_{mid} N).$$

We begin with the construction of the map

$$\omega(K) \otimes \omega(N) \to \omega(K \star_! N).$$

Let us define

$$\mathcal{K} := j_{0!}Rj_{\infty\star}K, \quad \mathcal{N} := j_{0!}Rj_{\infty\star}K.$$

Then we have

$$\omega(K) = H^\star(\mathbb{P}^1, \mathcal{K}), \quad \omega(N) = H^\star(\mathbb{P}^1, \mathcal{N}),$$

and hence we have

$$\omega(K) \otimes \omega(N) = H^\star(\mathbb{P}^1 \times \mathbb{P}^1, \mathcal{K} \boxtimes \mathcal{N}).$$

We would like to extend the multiplication map $\mathbb{G}_m \times \mathbb{G}_m \to \mathbb{G}_m$ to a map $\mathbb{P}^1 \times \mathbb{P}^1 \to \mathbb{P}^1$. We cannot do this, but if we omit from $\mathbb{P}^1 \times \mathbb{P}^1$ the two points $(0, \infty)$ and $(\infty, 0)$, then we can define a map

$$\pi : \mathbb{P}^1 \times \mathbb{P}^1 \setminus \{(0, \infty), (\infty, 0)\} \to \mathbb{P}^1,$$

given in homogeneous coordinates $(A, B), (X, Y)$ on $\mathbb{P}^1 \times \mathbb{P}^1$ by

$$((A, B), (X, Y)) \mapsto (AX, BY).$$

Over the open set $\mathbb{G}_m$ in the target, we have the multiplication map $\mathbb{G}_m \times \mathbb{G}_m \to \mathbb{G}_m$. Over the point 0 in the target, the fibre $\pi^{-1}(0)$ is the union of $0 \times \mathbb{A}^1$ with $\mathbb{A}^1 \times 0$. Notice that because both $\mathcal{K}$ and $\mathcal{N}$ vanish at 0, the restriction of $\mathcal{K} \boxtimes \mathcal{N}$ to $\pi^{-1}(0)$ vanishes. [The fibre $\pi^{-1}(\infty)$ is the union of $\infty \times (\mathbb{P}^1 \setminus 0)$ with $(\mathbb{P}^1 \setminus 0) \times \infty$, a fact we will use later.]

Because $\mathcal{K} \boxtimes \mathcal{N}$ vanishes at both the points $(0, \infty)$ and $(\infty, 0)$, we have an isomorphism

$$H_c^\star(\mathbb{P}^1 \times \mathbb{P}^1 \setminus \{(0, \infty), (\infty, 0)\}, \mathcal{K} \boxtimes \mathcal{N}) \cong H^\star(\mathbb{P}^1 \times \mathbb{P}^1, \mathcal{K} \boxtimes \mathcal{N}).$$

Slightly less obvious is the following lemma.

Lemma 30.5. *The restriction map in ordinary cohomology gives an isomorphism*

$$H^\star(\mathbb{P}^1 \times \mathbb{P}^1, \mathcal{K} \boxtimes \mathcal{N}) \cong H^\star(\mathbb{P}^1 \times \mathbb{P}^1 \setminus \{(0, \infty), (\infty, 0)\}, \mathcal{K} \boxtimes \mathcal{N}).$$

Proof. It suffices to show that, denoting by $i_{(0,\infty)}$ and $i_{(\infty,0)}$ the inclusions, we have

$$Ri^!_{(0,\infty)}(\mathcal{K} \boxtimes \mathcal{N}) = 0, \quad Ri^!_{(\infty,0)}(\mathcal{K} \boxtimes \mathcal{N}) = 0.$$

By duality, this is equivalent to

$$i^\star_{(0,\infty)}(D\mathcal{K} \boxtimes D\mathcal{N}) = 0, \quad i^\star_{(\infty,0)}(D\mathcal{K} \boxtimes D\mathcal{N}) = 0.$$

But both $D\mathcal{K} = Rj_{0\star}j_{\infty!}D\mathcal{K}$ and $D\mathcal{N} = Rj_{0\star}j_{\infty!}D\mathcal{N}$ vanish at ∞, so their external product $D\mathcal{K} \boxtimes D\mathcal{N}$ vanishes at both the points $(0, \infty)$ and $(\infty, 0)$. ∎

So our situation now is that we have isomorphisms

$$H_c^\star(\mathbb{P}^1 \times \mathbb{P}^1 \setminus \{(0, \infty), (\infty, 0)\}, \mathcal{K} \boxtimes \mathcal{N}) \cong H^\star(\mathbb{P}^1 \times \mathbb{P}^1, \mathcal{K} \boxtimes \mathcal{N}) = \omega(K) \otimes \omega(N)$$

$$= H^\star(\mathbb{P}^1 \times \mathbb{P}^1, \mathcal{K} \boxtimes \mathcal{N}) \cong H^\star(\mathbb{P}^1 \times \mathbb{P}^1 \setminus \{(0, \infty), (\infty, 0)\}, \mathcal{K} \boxtimes \mathcal{N}),$$

in which the composite isomorphism is the "forget supports" map.

By the Leray spectral sequence for π in compact cohomology, we have

$$H_c^\star(\mathbb{P}^1 \times \mathbb{P}^1 \setminus \{(0, \infty), (\infty, 0)\}, \mathcal{K} \boxtimes \mathcal{N}) = H^\star(\mathbb{P}^1, R\pi_!(\mathcal{K} \boxtimes \mathcal{N})).$$

And by the Leray spectral sequence in ordinary cohomology, we have

$$H^\star(\mathbb{P}^1 \times \mathbb{P}^1 \setminus \{(0,\infty),(\infty,0)\}, \mathcal{K} \boxtimes \mathcal{N}) \cong H^\star(\mathbb{P}^1, R\pi_\star(\mathcal{K} \boxtimes \mathcal{N})).$$

So we have isomorphisms

$$H^\star(\mathbb{P}^1, R\pi_!(\mathcal{K} \boxtimes \mathcal{N})) \cong \omega(K) \otimes \omega(N) \cong H^\star(\mathbb{P}^1, R\pi_\star(\mathcal{K} \boxtimes \mathcal{N})),$$

in which the composite map is induced by the "forget supports" map $R\pi_!(\mathcal{K} \boxtimes \mathcal{N}) \to R\pi_\star(\mathcal{K} \boxtimes \mathcal{N})$.

There are two further isomorphisms we now take into account. As already observed above, $\mathcal{K} \boxtimes \mathcal{N}$ vanishes on the fibre $\pi^{-1}(0)$, so by proper base change $R\pi_!(\mathcal{K}\boxtimes\mathcal{N})$ vanishes at 0, and hence the adjunction map $j_{0!}j_0^\star \to id$ induces an isomorphism

$$j_{0!}j_0^\star R\pi_!(\mathcal{K} \boxtimes \mathcal{N}) \cong R\pi_!(\mathcal{K} \boxtimes \mathcal{N}).$$

We claim that, dually, the adjunction map $id \to Rj_{\infty\star}j_\infty^\star$ induces an isomorphism

$$R\pi_\star(\mathcal{K} \boxtimes \mathcal{N}) \cong Rj_{\infty\star}j_\infty^\star R\pi_\star(\mathcal{K} \boxtimes \mathcal{N}).$$

Indeed, this is equivalent, by duality, to the statement that the adjunction map $j_{\infty!}j_\infty^\star \to id$ induces an isomorphism

$$j_{\infty!}j_\infty^\star R\pi_!(D\mathcal{K} \boxtimes D\mathcal{N}) \cong R\pi_!(D\mathcal{K} \boxtimes D\mathcal{N}).$$

This holds, because both $D\mathcal{K}$ and $D\mathcal{N}$ vanish at ∞, so their external product $D\mathcal{K} \boxtimes D\mathcal{N}$ vanishes on $\pi^{-1}(\infty)$ (which we noted above is the union of $\infty \times (\mathbb{P}^1 \setminus 0)$ with $(\mathbb{P}^1 \setminus 0) \times \infty$), and we get the assertion by proper base change.

Applying the adjunction map $id \to Rj_{\infty\star}j_\infty^\star$, we get a map

$$j_{0!}j_0^\star R\pi_!(\mathcal{K} \boxtimes \mathcal{N}) \to Rj_{\infty\star}j_\infty^\star j_{0!}j_0^\star R\pi_!(\mathcal{K} \boxtimes \mathcal{N}).$$

Here the cohomology of the target is $\omega(K \star_! N)$.

Applying the adjunction map $j_{0!}j_0^\star \to id$, we get a map

$$j_{0!}j_0^\star j_{\infty!}j_\infty^\star R\pi_\star(D\mathcal{K} \boxtimes D\mathcal{N}) \to j_{\infty!}j_\infty^\star R\pi_\star(D\mathcal{K} \boxtimes D\mathcal{N}).$$

Here the cohomology of the target is $\omega(K \star_\star N)$.

So the situation now is that we have a diagram of horizontal "forget supports" maps and vertical adjunction maps as follows. To make the diagram fit horizontally on the page, we write S for the subset $\{(0,\infty),(\infty,0)\}$ of $\mathbb{P}^1 \times \mathbb{P}^1$. Thus

$$\mathbb{P}^1 \times \mathbb{P}^1 \setminus S := \mathbb{P}^1 \times \mathbb{P}^1 \setminus \{(0,\infty),(\infty,0)\},$$

and our diagram is this.

$$\begin{array}{ccc}
\omega(K) \otimes \omega(N) & \xrightarrow{\ =\ } & \omega(K) \otimes \omega(N) \\
\Big\downarrow{\scriptstyle =} & & \Big\downarrow{\scriptstyle =} \\
H^\star(\mathbb{P}^1 \times \mathbb{P}^1, \mathcal{K} \boxtimes \mathcal{N}) & \xrightarrow{\ =\ } & H^\star(\mathbb{P}^1 \times \mathbb{P}^1, \mathcal{K} \boxtimes \mathcal{N}) \\
\Big\uparrow{\scriptstyle \cong} & & \Big\downarrow{\scriptstyle \cong} \\
H_c^\star(\mathbb{P}^1 \times \mathbb{P}^1 \setminus S, \mathcal{K} \boxtimes \mathcal{N}) & \xrightarrow{\ \cong\ } & H^\star(\mathbb{P}^1 \times \mathbb{P}^1 \setminus S, \mathcal{K} \boxtimes \mathcal{N}) \\
\Big\downarrow{\scriptstyle \cong,\ Leray} & & \Big\downarrow{\scriptstyle \cong,\ Leray} \\
H^\star(\mathbb{P}^1, R\pi_!(\mathcal{K} \boxtimes \mathcal{N})) & \xrightarrow{\ \cong\ } & H^\star(\mathbb{P}^1, R\pi_\star(\mathcal{K} \boxtimes \mathcal{N})) \\
\Big\uparrow{\scriptstyle \cong} & & \Big\downarrow{\scriptstyle \cong} \\
H^\star(\mathbb{P}^1, j_{0!}j_0^\star R\pi_!(\mathcal{K} \boxtimes \mathcal{N})) & & H^\star(\mathbb{P}^1, Rj_{\infty\star}j_\infty^\star R\pi_\star(\mathcal{K} \boxtimes \mathcal{N})) \\
\Big\downarrow & & \Big\uparrow \\
\end{array}$$

$$H^\star(\mathbb{P}^1, j_{0!}j_0^\star Rj_{\infty\star}j_\infty^\star R\pi_!(\mathcal{K} \boxtimes \mathcal{N})) \to H^\star(\mathbb{P}^1, j_{0!}j_0^\star Rj_{\infty\star}j_\infty^\star R\pi_\star(\mathcal{K} \boxtimes \mathcal{N}))$$

$$\begin{array}{ccc}
\Big\downarrow{\scriptstyle :=} & & \Big\uparrow{\scriptstyle :=} \\
\omega(K \star_! N) & \xrightarrow{\ \cong\ } & \omega(K \star_\star N)
\end{array}$$

So what must be shown is the commutativity of the following diagram.

$$\begin{array}{ccc}
H^\star(\mathbb{P}^1, R\pi_!(\mathcal{K} \boxtimes \mathcal{N})) & \xrightarrow{\ \cong\ } & H^\star(\mathbb{P}^1, R\pi_\star(\mathcal{K} \boxtimes \mathcal{N})) \\
\Big\uparrow{\scriptstyle \cong} & & \Big\downarrow{\scriptstyle \cong} \\
H^\star(\mathbb{P}^1, j_{0!}j_0^\star R\pi_!(\mathcal{K} \boxtimes \mathcal{N})) & & H^\star(\mathbb{P}^1, Rj_{\infty\star}j_\infty^\star R\pi_\star(\mathcal{K} \boxtimes \mathcal{N})) \\
\Big\downarrow & & \Big\uparrow \\
\end{array}$$

$$H^\star(\mathbb{P}^1, j_{0!}j_0^\star Rj_{\infty\star}j_\infty^\star R\pi_!(\mathcal{K} \boxtimes \mathcal{N})) \to H^\star(\mathbb{P}^1, j_{0!}j_0^\star Rj_{\infty\star}j_\infty^\star R\pi_\star(\mathcal{K} \boxtimes \mathcal{N}))$$

To see what is going on, let us define

$$\mathcal{A} := R\pi_!(\mathcal{K} \boxtimes \mathcal{N}), \quad \mathcal{B} := R\pi_\star(\mathcal{K} \boxtimes \mathcal{N}),$$

and denote by $f : \mathcal{A} \to \mathcal{B}$ the "forget supports" map. Then the diagram in question is gotten by applying the functor $\mathcal{C} \mapsto H^\star(\mathbb{P}^1, \mathcal{C})$ to the following diagram.

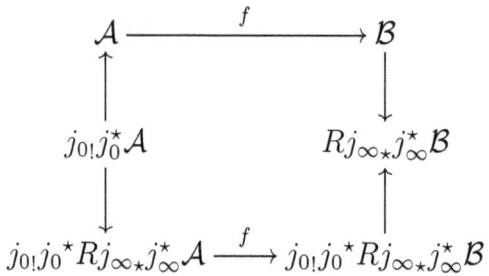

It suffices to show this this last diagram is commutative. To see this, we embed into the larger diagram, in which all the vertical and inward facing arrows are adjunctions, and all the horizontal arrows are induced by f.

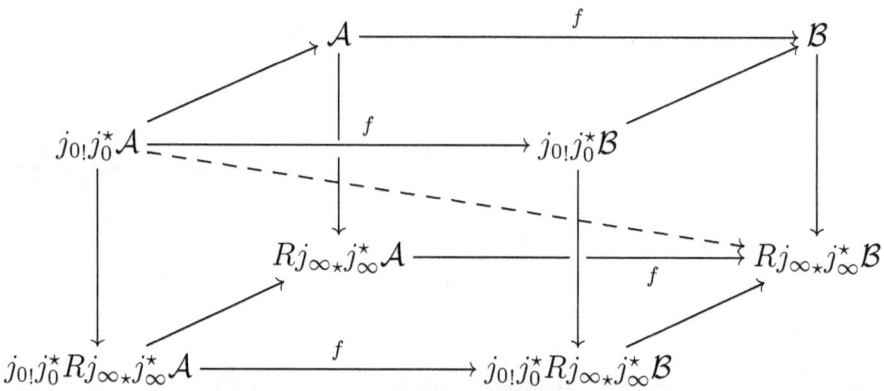

A diagram chase shows that it suffices to show that the top face, the right side face, and the front face are each commutative. The top face is obtained by applying the adjunction $j_{0!}j_0^\star \to id$ to the morphism f; it is commutative by the functoriality of adjunction. The right side face is obtained by applying the adjunction $id \to Rj_{\infty\star}j_\infty^\star$ to the (adjunction) morphism $j_{0!}j_0^\star\mathcal{B} \to \mathcal{B}$, so is commutative by the functoriality of adjunction. And the front face is obtained by applying the same adjunction to the morphism $j_{0!}j_0^\star(f)$, so is again commutative. [In fact, all six faces are commutative, always by the functoriality of adjunction.] This concludes the proof of Theorem 3.1.

Bibliography

[BBD] Beilinson, A., Bernstein, J., and Deligne, P., Faisceaux pervers. (entire contents of) Analyse et topologie sur les éspaces singuliers, I (Conférence de Luminy, 1981), 5-171, Astérisque, 100, Soc. Math. France, Paris, 1982.

[Bry] Brylinski, J.-L., Théorie du corps de classes de Kato et revêtements abéliens de surfaces. Annales de l'institut Fourier 33 (1983), n° 3 , 23-38.

[Chev-TGL II] Chevalley, C., Théorie des Groupes de Lie. tome II. Groupes Algêbriques. Actualitês Sci. Ind. no. 1152. Hermann & Cie., Paris, 1951. vii+189 pp.

[Dav] Davenport, H., On the distribution of quadratic residues (mod p). J. London Math. Soc. 6 (1931), 49-54, reprinted in The Collected Works of Harold Davenport (ed. Birch, Halberstam, Rogers), Academic Press, London, New York and San Francisco, 1977, Vol. IV, 1451-1456.

[D-H] Davenport, H., and Hasse, H., Die Nullstellen der Kongruenz-zetafunktionen in gewissen zkylishen Fallen. J. Reine Angew. Math. 172 (1934), 152-182, reprinted in The Collected Works of Harold Davenport (cf. [Dav]), Vol. IV, 188-1519.

[De-Const] Deligne, P., Les constantes des quations fonctionnelles des fonctions L. Modular functions of one variable, II (Proc. Internat. Summer School, Univ. Antwerp, Antwerp, 1972), pp. 501-597. Lecture Notes in Math., Vol. 349, Springer, Berlin, 1973.

[De-ST] Deligne, P., Applications de la formule des traces aux sommes trigonométriques. pp. 168-232 in SGA 4 1/2, cited below.

[De-Weil II] Deligne, P., La conjecture de Weil II. Publ. Math. IHES 52 (1981), 313-428.

[Fuj-Indep] Fujiwara, K., Independence of ℓ for intersection cohomology (after Gabber). Algebraic Geometry 2000, Azumino (Hotaka), 141-151, Adv. Stud. Pure Math., 36, Math. Soc. Japan, Tokyo, 2002.

[Ga-Loe] Gabber, O, and Loeser, F., Faisceaux pervers l-adiques sur un tore. Duke Math. J. 83 (1996), no. 3, 501-606.

[GGS] Goldstein, D., Guralnick, R., and Strong, R., A lower bound for the dimension of a highest weight module, to appear.

[Gr-Rat] Grothendieck, A., Formule de Lefschetz et rationalité des fonctions L. Séminaire Bourbaki, Vol. 9, Exp. No. 279, 41-55, Soc. Math. France, 1995.

[Ha-Ell] Hasse, H., Zur Theorie der abstrakten elliptischen Funktionenkörper I, II, III, J. Reine Angew. Math. 175 (1936), 55-62, 69-88, and 193-208.

[Ha-Rel] Hasse, H., Theorie der relativ-zyklischen algebraischen Funktionenkörper, insbesondere bie endlichen Konstarntkörper, J. Reine Angew. Math. 172 (1934), 37-54.

[Ka-ACT] Katz, N., Affine Cohomological Transforms, Perversity, and Monodromy. J. Amer. Math. Soc. 6 (1993), no. 1, 149-222.

[Ka-ClausCar] Katz, N., From Clausen to Carlitz: Low-dimensional spin groups and identities among character sums. Moscow Math. J. 9 (2009), no. 1, 57-89.

[Ka-ESDE] Katz, N., Exponential sums and differential equations. Annals of Mathematics Studies, 124. Princeton Univ. Press, Princeton, NJ, 1990. xii+430 pp.

[Ka-G2Hyper] Katz, N., G_2 and hypergeometric sheaves. Finite Fields Appl. 13 (2007), no. 2, pp. 175-223.

[Ka-GKM] Katz, N., Gauss sums, Kloosterman sums, and monodromy groups. Annals of Math. Studies, 116. Princeton Univ. Press, Princeton, NJ,1988. x+246 pp.

[Ka-Lau] Katz, N., and Laumon, G., Transformation de Fourier et majoration de sommes exponentielles. Pub. Math. IHES 62 (1985), 361-418.

[Ka-MMP] Katz, N., Moments, monodromy, and perversity: a Diophantine perspective. Annals of Mathematics Studies, 159. Princeton University Press, Princeton, NJ, 2005. viii+475 pp.

[Ka-NotesG2] Katz, N., Notes on G_2, determinants, and equidistribution, Finite Fields and their Applications 10 (2004), 221-269.

[Ka-RLS] Katz, N., Rigid local systems. Annals of Mathematics Studies, 139. Princeton University Press, Princeton, NJ, 1996. viii+223 pp.

[Ka-Sar] Katz, N., and Sarnak, P., Random matrices, Frobenius eigenvalues, and monodromy. American Mathematical Society Colloquium Publications, 45. American Mathematical Society, Providence, RI, 1999. xii+419 pp.

[Ka-SE] Katz, N., Sommes exponentielles. [Exponential sums] Course taught at the University of Paris, Orsay, Fall 1979. With a preface by Luc Illusie. Notes written by Gérard Laumon. With an English summary. Astérisque, 79. Société Mathématique de France, Paris, 1980. 209 pp.

[Kloos] Kloosterman, H. D., On the representation of numbers in the form $ax^2 + by^2 + cz^2 + dt^2$. Acta Mathematica 49 (1926), 407-464.

[KRR] Kurlberg, P., Rosenzweig, L., and Rudnick, Z., Matrix elements for the quantum cat map: fluctuations in short windows. Nonlinearity 20 (2007), no. 10, 2289-2304.

[Lau-CC] Laumon, G., Comparaison de caractéristiques d'Euler-Poincaré en cohomologie l-adique. C. R. Acad. Sci. Paris Sér. I Math. 292 (1981), no. 3, 209-212.

[Lau-SCCS] Laumon, G., Semi-continuité du conducteur de Swan (d'après P. Deligne). Caractéristique d'Euler-Poincaré, pp. 173-219, Astérisque, 82-83, Soc. Math. France, Paris, 1981.

[Lau-TFCEF] Laumon, G., Transformation de Fourier, constantes d'équations fonctionnelles et conjecture de Weil. Pub. Math. IHES 65 (1987), 131-210.

[Lu] Lübeck, F., Small degree representations of finite Chevalley groups in defining characteristic. LMS J. Comput. Math. 4 (2001), 135-169 (electronic).

[Ray] Raynaud, M., Caractéristique d'Euler-Poincaré d'un faisceau et cohomologie des variétés abéliennes. Séminaire Bourbaki, Vol. 9, Exp. No. 286, 129-147, Soc. Math. France, Paris, 1995.

[Ru] Rudin, W., Real and Complex Analysis, McGraw-Hill, New York, 1987. xiv+416 pp.

[Se-Dri] Serre, J.-P., Sur les groupes de Galois attachés aux groupes p-divisibles. Proc. Conf. Local Fields (Driebergen, 1966), pp. 118-131, Springer, Berlin, 1967.

[Se-Let] Serre, J.-P., Lettre à Ken Ribet du 1/1/1981. Oeuvres. IV. 1985-1998, pp. 1-17, Springer, Berlin, 2000.

[Se-Rep] Serre, J.-P., Représentations ℓ-adiques. Algebraic Number Theory (Kyoto Internat. Sympos., Res. Inst. Math. Sci., Univ. Kyoto, Kyoto, 1976) pp. 177-193, Jap. Soc. Prom. Sci., Tokyo, 1977, reprinted in Oeuvres. Vol. III. 1972-1984, Springer, Berlin, 1986, 384-400.

[SGA 4 1/2] Cohomologie Etale. Séminaire de Géométrie Algébrique du Bois Marie SGA 4 1/2. par P. Deligne, avec la collaboration de J. F. Boutot, A. Grothendieck, L. Illusie, et J. L. Verdier. Lecture Notes in Mathematics, Vol. 569, Springer-Verlag, 1977.

[Weil] Weil, Andé, Variétés abéliennes et courbes algébriques. Actualités Sci. Ind., no. 1064 = Publ. Inst. Math. Univ. Strasbourg 8 (1946). Hermann & Cie., Paris, 1948. 165 pp.

Index

GPSR Authorized Representative: Easy Access System Europe - Mustamäe tee 50, 10621 Tallinn, Estonia, gpsr.requests@easproject.com